AF862358

THE BIOLOGY OF AGEING

Symposia of the Institute of Biology

BIOLOGICAL HAZARDS OF ATOMIC ENERGY
Oxford University Press. 1952

FREEZING AND DRYING
Institute of Biology 1952

THE BIOLOGY OF DESERTS
Institute of Biology 1954

THE NUMBERS OF MAN AND ANIMALS
Oliver and Boyd 1955

BIOLOGICAL ASPECTS OF THE TRANSMISSION OF DISEASE
Oliver and Boyd 1957

THE BIOLOGY OF AGEING

Symposia of the Institute of Biology, No. 6

EDITED BY

W. B. YAPP

and

G. H. BOURNE

LOS ALAMOS
SCIENTIFIC LABORATORY
WITHDRAWN
OCT 29 1957
LIBRARIES
PROPERTY

Hafner Publishing Company
New York

Printed in England

Church Army Press, Cowley, Oxford, England 3323

FOREWORD

THE Institute of Biology has from time to time organized symposia, and the papers read at five of these have now been published. The subjects have, by design, been somewhat broader than those which are dealt with by the existing specialist societies or learned journals, and some of them have included contributions from speakers who would hardly, in the ordinary sense of the word, be looked on as biologists. One result of this has been to make it difficult for librarians to catalogue the published volumes, and the five that have already appeared are widely scattered on the shelves. It would seem to be easier for readers if all the symposia were brought together, and this would also emphasize, what sometimes needs enphasizing, that Biology is a broad discipline which can help in understanding many aspects in the life of man. The Council of the Institute has accordingly decided that the symposia shall take the form of a journal, with the volumes to appear as far as possible annually. *The Biology of Ageing*, the volume for 1957, becomes number 6, and the titles of the earlier volumes are set out on p. ii. The series may be quoted as " *Symp. Inst. Biol.*".

The present symposium was set in galley proof before the new editorial policy was decided, and we apologize to readers for some lack of uniformity in the volume and particularly in the form of the references. To have brought all the contributions into line would have delayed publication unduly.

W. B. YAPP,
Editor for Symposia.

ACKNOWLEDGEMENTS

This volume is a record of the Symposium held at University College, London, on the 27th and 28th September, 1956. We thank all those who helped in any way to make the meeting possible, and especially the contributors; the Chairmen of the four sessions (Professor G. P. Wells, Dr. A. S. Parkes, Mr. W. A. Sanderson and Professor W. J. E. Jessop); University College, London, for the use of the lecture theatre; and the Nuffield Foundation for a grant which enabled the papers to be printed and circulated before the meeting.

CONTENTS

INTRODUCTION

By

F. LE GROS CLARK AND N. W. PIRIE

A SYMPOSIUM on ageing needs some introduction. The subject has about it a deceptive simplicity, but that is partly because we have tended to assume that ageing as we know it in man must in some way characterize any plant or animal that has lived long enough. Only of late years has the subject received much attention from research departments—and that for reasons that are all too human.

Societies, such as our own, that contain a steadily increasing number of pensioners, were bound to grow more interested in the nature of organic ageing and to ask whether its onset could be delayed. When biologists tried to oblige by providing a few of the answers, they usually encountered certain difficulties, which are probably inherent in the subject itself; senescence primarily means and always has meant *human* senescence, and human senescence is a topic that traditionally carries with it various emotional and literary overtones. Since even a biologist cannot wholly escape the influence that these overtones may have upon his scientific judgment, his only safe research procedure is to make what allowances he can for them.

A comment should be added on the terminology of the subject. The contributors to the symposium have been reasonably consistent in their usage, but the terms are popularly as well as scientifically used, and they appear to be in some confusion. What, for example, do we mean by the word *ageing*? The trouble is that for the layman it almost invariably suggests the onset of true old age, whereas the more precise biologist may argue that it ought to include the whole sequence of changes that start with the birth or even the conception of an organism. The term " senescence " is equally open to misinterpretation. By its derivation and classical usage it should mean the process of becoming old, yet it seems frequently to be used by the layman, and possibly at times by the scientist, as though it implied a fixed state or prolonged phase in the life of an organism. Before the biological sciences go much further into the problem, they would be well advised to standardize the vocabulary.

Familiar as it is to all of us, the term ageing needs some defining. The process that it represents cannot be a matter of time alone, but must be describable in terms of organic changes of some kind. In the long run living things will all die, even if they do not age, because they are all exposed to risks that may at any time prove lethal. Ageing must mean that organisms (though perhaps not all organisms) come to experience in the course of time a gradually *increasing* liability to die, and often therefore to die from causes that might formerly have left them more or less unaffected. An organism that ages in this sense does not normally encounter such a decline in its viability until it has lived for a certain period of time. While we may recognize or think we recognize such a senescent process at work in the members of many species of animals, we must not take for granted that the same deteriorative factors are at work in all of them. Indeed, to look for a cause of ageing common to all living things may be quite abortive. For one thing, it seems very probable that many organisms cannot be described as ageing at all. Where ageing does occur in living things it should perhaps be treated, as Dr. Comfort has suggested, rather as a unity of effects than as a unity of causes. The factors that are operating may turn out to be too complex and too varied to allow of any neat biological classification.

The determination of the vulnerability of the members of an animal or plant population, with the passage of time, would be one way of measuring and comparing types of ageing. All we should need to know would be the chronological ages of all the individuals at risk within our sample. They have, of course, to succumb before we know whether and at what rate their chances of dying have increased with their chronological ages. It has been more robustly suggested in the symposium that another method of tracing deteriorative changes could be adopted—that of measuring the working vigour of men and of some of their domesticated animals against the demands of fairly constant day-to-day tasks. If life tables can be constructed, why not "working-life tables"? The time of life at which working men or draught animals can no longer carry out their tasks in the customary way surely marks a definite state in the ageing process. The method would, however, be of limited statistical application and would involve considerably more guess work than a calculation based on mortality rates.

A common-sense objection to measuring the degree of ageing in either of these ways is that they would often give a minimum not at birth but at some later age. For man, for birds and for many other animals the chances of dying are high in early life, and then fall to a minimum in what, in human terms, may be called late adolescence. There would be a similar, though less precise, maximum expectation of working life. It is a matter for discussion and research to decide whether these facts would constitute a biological objection to a definition of ageing made in this way.

Since there is a distinctly human quality in the common ideas we have about senescence, several of the contributions to the symposium were confined to human aspects of the problem. It would indeed be difficult to discuss the papers as a whole without making continual reference to the human implications of the subject. The contributions provide us with a brief survey of the biological problem of ageing, as it is at present understood. Obviously what the writers say is not all that could have been said upon the subject, but they do indicate many of the biological questions that have now to be asked and if possible answered.

The first few papers illustrate clearly our need to distinguish the senile changes that may take place in a cell from those that may be characteristic of tissues in a living body or of a whole many-celled organism. Sir Cyril Hinshelwood deals only with single-celled organisms such as bacteria. He points out that a cell is not static, but is a system through which are flowing the materials it metabolizes; it is therefore being constantly replaced in all its parts. It has, one might say, no more permanence than the vortex in an endlessly emptying bath that is being as endlessly replenished. He interprets ageing in such an organism as a failure to make the absorption and retention of metabolites keep pace with losses due to leakage; the deficiency could occur either through a reduced absorption or through an increased leakiness. In the long run failures of this kind would make a cell unworkable, and he prefers such an interpretation to one that is based on the theory of the inactivation of some single and essential mechanism in the cell.

When many cells are integrated into an organism, the position is not necessarily very different, though when one comes to some of the higher animals it may, of course, differ markedly. Dr. Heath likens a long-lived tree to a frame of material that is no longer living, on which there hangs, as it were, a sheath of cells that are constantly being renewed. A Sequoia three hundred feet high that has lived for three thousand years may contain no living cell that is more than a few years old. It bears a closer resemblance to a sponge-colony than to an elephant or a man. The whole system can survive (like one of G. B. Shaw's " ancients ") until it meets its inevitable destructive accident in tornado, lightning or forest fire.

Annual plants age or appear to age in a more obvious way, and this has often been associated by observers with such processes as flowering and fruiting. Dr. Heath, however, quotes experiments in which flowering has been stopped without any effect upon the manifestations usually interpreted as old age in the plants. We have had here incidentally what appears to be an attempt on the part of some writers to graft a perennial illusion on to an annual plant. Many philosophers had assumed that sexual manifestations promote ageing or inevitably precede its onset. This assumption runs through much of mediaeval and Calvinist thinking; and with Donne it even reached the dignity of poetry:

> " Since each such act they say
> Diminisheth the length of life a day."

But the Taoists of China in the sixth century A.D. adopted the more cheerful supposition that sexual activity tends to delay senescence.

Apparently this belief in the sex-influenced contraction of life, whether plant or human, was based less on any moral principle than on the idea that an organism must at last run through its allotted reserves of energy or lose the structural capacity for dealing with the energy it so gloriously expends. The idea that an organism lives till it has metabolized a definite amount of bodily energy is only a more generalized version of the same theme. Rubner, for instance, once surmised that the amount would ordinarily be about 200,000 kilocalories for every kilogram of an animal's adult body weight; he placed man in the exceptional position of being able to metabolize about four times as much as this. If that were so, man's place in nature would take some explaining. But Professor Bourlière shows below that things are not quite as simple as this, and that several other animals are also exceptional. It may be true that the " rate of living " of a species does in some way determine its average span of life, and that, for example, mammals that periodically reduce their metabolic rate by hibernation tend to live longer than non-hibernating mammals of similar size. Yet precisely what is being used up or exhausted, it would be difficult to say. We do at least know that it is not necessarily the people who enjoy long periods of lethargy that live longest; but it may be quite unjustifiable to compare them with bats.

Dr. Bourne reverts to the problem of the ageing of cells, this time as studied histologically within a multicellular organism. He describes changes in the mitochondria of cells, which in the main seem to result in an increased fragility; the cell, like the whole organism, finds, as it were, that its parts break more easily and repair less easily as it grows chronologically older. There are also more clearly definable changes going on within it, such as the accumulation of calcium, iron and pigments, and a variation in the concentrations of many of the enzymes. The effects of ageing in an animal's tissues are not readily studied, because in many tissues the individual cells, like the cells of a tree, are being constantly renewed. But in tissues that experience no cell division of this kind there is a steady loss of cells, and this is perhaps most clearly seen in the central nervous system. Old age may, in fact, be due to an ultimately suicidal activity of the cells themselves. We can agree that it is tempting to throw much of the blame for senescence upon the cells of which we are indubitably composed, but it is unlikely that in the higher organisms the process of ageing could thus be reduced to a single uniform causal sequence of events. Cellular changes of this kind will, however, be investigated with growing interest in years to come.

Changes in the hormones with age remind us again of the old idea that sexuality and age are somehow related. Because for many people the most exasperating feature of old age is apt to be the decay of sexual capacity and enthusiasm, there is a tendency to overlook the fact that eunuchs often live to a ripe old age. The eunuch Narses was well beyond seventy when he won his final battle against the Goths, and almost ninety before he was retired from the governorship of Italy. Dr Swyer surveys our knowledge of the changes in hormone levels throughout the course of ageing, and describes the attempts that have been made to circumvent them. The old techniques of modifying or transplanting organs such as testes do not now find favour, but the administration of suitable mixtures of the steroid hormones that become deficient in old age seems to be effective in preventing or counteracting some at least of the symptoms of advanced senescence. It is much to be hoped that this line of research will be diligently pursued; it is deplorable (as was remarked by some speakers in the discussion) that a disproportionate amount of attention is given to the effects of old age on men; for women, though they tend to live longer than men, appear to suffer more from the disabilities than might well be remediable by this type of replacement therapy.

The notion that certain properties in a man's diet could conceivably prolong his life or vigour is no less ancient. Dr. Sinclair suggests reasons for doubt. In common with several other contributors he recalls the evidence that to force the rate of growth in a young animal may shorten its expectation of life. An animal's food is, of course, important in the sense that throughout its life chemical interchanges are always taking place. The structure and conformation of the body of a man of sixty depend to some extent on the quality and quantity of the food to which he has been accustomed. It is probably easy enough for him to arrange his diet so as to shorten his expectation of life; it is questionable whether a careful and ingenious planning of his diet would have a greater effect in delaying the senescent process than would any normal mixed diet taken in moderation.

Those who feel that inheritance must somehow be playing a part in longevity will find in the paper by Mr. Maynard Smith an indication of the difficulties that attend research of this kind. The fact that under favourable conditions the members of one species apparently live much longer than members of another species of about the same size inevitably suggests the assumption that various selective forces must have been at work somewhere in their evolution, but what these selective forces were and how they worked is not at all clear. As Dr. Comfort points out, where the risks of death are numerous and constant, few if any of the members of a species might ever live long enough for senescence to manifest itself among them. This is possibly the fate of many small birds and mammals in the wild. Specific differences in longevity still remain, however, and they are not readily accounted for. It would be far simpler if we could fall back on the traditional belief that each species must have been allotted its appropriate life span through the inscrutable designs of Providence.

Dr. Comfort's paper on ageing in animals leads naturally to man. The impression we have of ageing in the animals with which we are most familiar is much the same as the impression of ageing as it is experienced by the members of our own species. Whether the complex of causes is invariably the same is another matter. We shall plainly not get much further until we possess detailed life tables for a number of species. It has been suggested in the course of the symposium that comparative information might be obtained by observations on the working lives of such draught animals as horses and elephants, but doubt is expressed whether we have not come too late in industrial history to hope for reliable records. The possible use of the Burma forest elephant as a laboratory animal in the study of ageing offers some exciting prospects.

In his survey of the clinical aspects of senescence Professor Hobson brings into relief the fact that old age is frequently accompanied by chronic diseases and ailments of one kind or another. Dr. Heron emphasizes that in man at all events age is complicated by many psychological changes. Mr. Benjamin completes the human picture by displaying the vast and shifting demographic framework within which the drama of ageing is now being played out in our own country.

Our approach to the problems of senescence must necessarily be in some measure subjective, as indeed are the conceptions we tend to adopt about our own place in society when we arrive at a chronologically advanced age. How subjective our approach is, we can scarcely realize, so ingrained have become our traditional habits of thought. Most of us have our words and gestures perfect as we live through the Shakespearean Seven Ages of Man. Yet a biological residue remains which is clearly entitled to be called organic ageing. What precisely accumulates or breaks down or fails to function or runs out or goes to seed, we do not know, and it may be that none of these metaphors is strictly applicable. We do not know whether we shall later find ourselves describing these experiences in neurological or biochemical or cytological terms; the chances are that no one of the biological sciences will alone prove adequate to our purpose.

AGEING IN BACTERIA

By

C. N. Hinshelwood

Physical Chemistry Laboratory, University of Oxford

When the full rhythm of its existence is established and maintained a unicellular organism increases each component of its structure so as to preserve constant ratios of them all. At a certain point, probably largely determined by the deoxyribose nucleic acid content of the cell, division takes place in bacteria by binary fission. The number of cells, and the masses of the various components, all increase according to an exponential law, and with the same rate constant for each.

$$n_t = n_0 e^{kt},\ (m_1)t = (m_1)_0 e^{kt} \text{ and so on.}$$

The steady rhythm of growth and reproduction is seldom maintained for long, and indeed with cells such as bacteria, can only continue under rather carefully devised laboratory conditions. If one or other essential nutrient becomes exhausted growth is thereby halted. Otherwise, metabolic products, some of which are potentially toxic, accumulate, and growth is limited by their action. The development of an adverse *pH* is in fact one of such influences. When bacteria are grown in ordinary laboratory cultures the one or the other of these causes may be the limiting factor according to circumstances.

When growth is interrupted and the cells enter a non-proliferating phase the nicely balanced co-ordination which characterized the phase of exponential increase is upset. To understand the ways in which it is disturbed some attention must first be given to the kind of harmony which had been established during the preceding steady state.

Compounds containing the various essential elements are taken in by cells from the medium and initiate a complex series of reactions in which the cell material is reproduced. For the immense variety of chemical syntheses there must be many sequences of consecutive reactions. These sequences must branch and interlock in a complicated scheme which constitutes the reaction pattern of the cell.

It is very unlikely indeed that there are individual substances capable of autosynthesis in isolation and in their own right. The process of self-duplication is much more probably the result of a co-ordinated interplay of reactions in which the products of one set of enzymes or cell constituents build up the material of another, and so on in an interlocking system according to equations of the type

$$dX_1/dt=a_1X_2,\ dX_2/dt=a_2X_3 \text{ and so on.}$$

It is easy to show that in the steady state of such a system $dX_1/dt=kX_1$, $dX_2/dt=kX_2$ and so on, all the k's being equal, and the ratios X_1/X_2, X_1/X_3 . . . settling down to standard average values. These ratios are maintained as long as logarithmic growth goes on in a constant medium. If the cells are transferred to a new medium where the relative rates of different individual reaction steps of the complex reaction pattern are changed, the cell automatically undergoes an adaptive modulation to a new set of ratios.

When the maintenance of the complete rhythm is interrupted by the failure of supplies, there is no necessity at all for all the individual reaction steps to cease simultaneously and they do not do so. After the fully co-ordinated working (which leads to increase in size followed by periodic division) has been upset many different things may happen to bacterial cells.

In some circumstances division may continue without increase in the total mass of the culture, the individual cells becoming smaller. When this occurs the amount of deoxyribose nucleic acid, in some examples at least, tends to remain at a constant level in each cell. Thus more is synthesized from somewhere. The necessary phosphorus has been shown in certain examples to be derived from the ribose nucleic acid already formed. Direct interconversion of ribose nucleic acid and deoxyribose nucleic acid being unlikely in the circumstances, a reversal of the condensation processes by which the ribose nucleic acid was originally formed seems to be involved. The nucleotides derived from this breakdown can then be used, for the synthesis of more deoxyribose nucleic acid, possibly after suffering further intermediate breakdown themselves. In other conditions certain labile phosphates in the cell can be shown to increase at the end of the lag phase to a steady level which is reached at the expense of inorganic phosphate in the medium and which is maintained during the logarithmic period. The level falls again as the stationary phase sets in, when synthesis of acid-stable phosphates continues for a time at the expense of the labile group.

Such observations on the behaviour of cells in conditions of phosphorus shortage exemplify the way in which reactions which had been in balance continue in a less co-ordinated way. Even more significantly they indicate how certain synthetic processes (e.g. the deoxyribose nucleic acid formation) can occur by a re-allocation of the internal resources of the cell, made possible by a reversal of previous syntheses. This principle will prove of importance in what follows.

The consumption of labile intermediates, the re-distribution of material by degradation and fresh synthesis, the decay of enzyme systems, and sometimes actual lysis of parts of the cell, are all processes which cause serious departures from the steady state set up in logarithmic growth. If they occur they have to be largely reversed before the logarithmic phase can once more be resumed. The necessary restoration of previous conditions occurs during the lag phase, which is the more prolonged the profounder the preceding disturbance has been.

Enzymes, of course, can function independently of the complete cell economy, and even in relatively pure isolated preparations. Non-dividing cells, however, can do more than single enzymes and although their reaction networks are not fully operational, can use substantial parts of these, not merely for the catalysing of individual chemical reactions like an enzyme preparation but for more elaborately co-ordinated activities.

The breakdown of glucose and other carbon compounds, so fundamental for the complete process of growth and multiplication. can of course be observed with suspensions of bacteria which are prevented from growing by the absence of any nitrogen source. The use of radioactive tracers shows that in the early stages of this process (with coliform bacteria) phosphate anions and potassium ions flow into the cell. The K^+ concentration in the cell rises to a maximum (which is the greater the more acid the medium) and falls again as the reaction comes to an end. If an assimilable nitrogen source is provided so that multiplication can go on, the potassium concentration remains at the high level, and indeed the growth does not take place unless the potassium (or a suitable alternative alkali cation) is available.

Even when the perfect pattern of its self-generating existence is upset, the cell still strives to observe, or rather cannot help observing, the principle of " business as usual " throughout as much as possible of its intricate little economy.

Studies on exchanges of radioactive substances between non-dividing cells and the surrounding medium throw further light on this aspect of the matter. Bacterial cultures can be prepared by growth in media containing radioactive sulphur, phosphorus or carbon. The cells containing this labelled material can then be suspended in a medium lacking one of the essential components for growth, and as they sojourn here any exchange with the medium can be followed.

In these circumstances what is observed is slow steady loss of the active element from the cell. The loss, however, becomes slower and practically ceases when the viable count of the suspension has fallen nearly to zero. The loss occurs in a reasonably rapid manner (over days) only while the cells are living, and when they are dead it stops even though there is still plenty of active material left in the cells. It is not due, therefore, to complete destruction or disintegration or lysis of the bacteria. The experiments can be inverted, with the active elements added to the external medium.

Little actual uptake of the labelled elements is then observed. What occurs is thus predominantly a slow one-way leakage but one which involves essentially the living cells. The explanation seems to be as follows. There is a lively internal interchange of materials within the living but non-dividing cell, compounds formed by reversed synthesis in one part being more or less avidly salved for re-synthesis in another (as in the example of the *RNA*→ *DNA* conversion). The salvage mechanism is, however, not perfectly efficient and some of the intermediates leak slowly out. The time scale of the leakage is such that the experiments extend over a period which may vary from several days to several weeks.

Since radioactive markers began to be available for the investigation of the detailed balance sheet of reactions in cells and tissues, the realization has been growing that the internal economy of living matter is more dynamic than had been assumed. Many constituents of the animal and vegetable organism, beneath an appearance of permanence, are dissolved and renewed according to a time scale of days or weeks. It is true that this almost transatlantic zest for scrapping and replacement is not universal, nor so great that appreciable exchanges can be detected during the period of active growth itself. Spiegelman, Monod and others find that proteins do not turn over to any important extent during the actual growth of cells of *Esch. coli*. Radioactive tracers show here that enzymes such as galactosidase are built up during growth in a nearly irreversible manner, that is to say that the observed net rate of formation is not the difference of a much larger gross rate of formation compensated by a substantial rate of reverse reaction.

But bacterial cells have generation times measured in minutes or hours, so that even if logarithmic growth effectively involves one-way assimilation the balance of synthesis and degradation is none the less important in the existence of non-dividing cells.

In this, to the external view, static condition, an exact balance is not easy to maintain. Eventually, unless the cell, as it were, advances, it must go back, and a phase of decline usually sets in. With some cells this occurs rapidly, as in the lysis of some strains of *Bacillus subtilis*. Bacterial spores achieve a relatively very stable non-dividing economy of low activity, but commonly the processes of decay and repair cannot cancel one another indefinitely.

Enzyme activities fall, irreversible losses of diffusible intermediates occur, lytic processes supervene. Death of the cell is the final result.

The process of ageing and death in a unicellular organism is one of great interest and the subject of much controversy. A large population of unicellular organisms declines with time according to a law which often approximates to the exponential form $n=n_0e^{-\lambda t}$, where n_0 is the initial number of living cells, n the number at time t and λ the decay constant, which naturally varies widely according to the state of the environment and the presence or absence of toxic substances. The exponential law is by no means exactly followed. Sometimes there is an initial period during which the death rate is quite low and it rises only after the adverse environment has had time to bring about progressive changes in the cells. Sometimes, on the other hand, there is a resistant tail of cells which survive much longer than the simple law would predict. But the exponential curve is often enough followed sufficiently closely for it to possess theoretical significance. At its face value such a law implies that the chance of death in a given short time interval for any cell is independent of its previous history, as in the well known phenomenon of the radioactive decay of atoms.

The interpretation given in the well known " target " theory is that some sensitive centre in the cell receives a random hit from a molecule of a toxic agent, or from a quantum of high energy radiation, and that the effect is lethal. Though it is just possible that an intensely localized destructive effect might be propagated by a chain reaction through a sufficient volume of the cell to disturb its economy, the target theory, at any rate as regards chemical agents, is inherently improbable and has frequently been criticized. Apart from the question of its inherent unlikeliness, the chief objections have been made on the ground that the exponential law is illusory and that adequate experimental technique would reveal an early region of the decay curve where the chance of death increases with time, as is required by theories that the toxic agents cause progressive damage. Nevertheless, there is no doubt that the exponential law is sometimes rather a good approximation. Therefore an interpretation of it should be sought which is free from the objections to the theory of random encounters. The view that the internal economy of a non-dividing cell still involves lively though imperfectly co-ordinated metabolic changes would provide just such an explanation.

In one sense there is a progressive decline when a cell is exposed to unfavourable conditions, but this need not be thought of as unidirectional and unopposed. Decay of enzymes will occur, actual lysis of localized parts of the material may set in, polymerization and condensation reactions may be reversed. But as some of these things happen, the internal medium of the cell is enriched by compounds containing carbon, nitrogen and other elements which can be used for the repair or expansion of other parts of the cell which can still function. The picture which presents itself is not of a uniform one-way dissolution and decline of all parts of the cell, but of incipient dissolution alternating with reconstruction. Breakdown fragments are not wholly wasted but often largely salved and utilized afresh. There will be decline but decline combated by a vigorous struggle for regeneration. Nevertheless, the co-ordination which maintains a harmonious ratio between all the constituents has been lost.

In these conditions there could be an actual waxing and waning in the amounts of various enzymes or in the intensity of various functions of the cell. Something analogous is *mutatis mutandis* observed in the actual growth of cells which are imperfectly adapted to the medium, bursts of rapid increase alternating with arrests in a most irregular way as one or other intermediate becomes available or runs out. In the declining economy the chaos can be even worse. Some parts of the cell may even acquire phage-like properties and prey upon neighbouring parts.

In such circumstances the waxing and waning of given pairs of enzymes, or types of cell material, might be likened to the changes of various animal species some of which prey upon others. In competitive systems of this kind, as Volterra showed, periodic changes in the numbers of the various species may be set up. When one species has plenty to prey on, it multiplies, and then consumes the prey so fast that this becomes scarce. Then there is a shortage of food for the predatory species which in turn declines while the prey has a fresh opportunity to increase.

There is a very good possibility that approximately, and for a certain period of time, the mutual relations of some important cell constituents can be described by equations similar to Volterra's, and that their amounts wax and wane like the numbers of the animal species in question. The oscillations set up will of course not be truly periodic with the cycles repeating themselves exactly, since this state of affairs would demand the fulfilment of rather precisely defined conditions, but a rough periodicity is quite plausible.

The periodicities will be multiple, applying to various pairs of cell units which can enter into relations analogous to those of predator and prey.

These various periodicities, moreover, will be largely independent of one another with their phases randomly related. In the declining cell economy there is no mechanism of adjustment, as there is in the gradual establishment of logarithmic growth, for harmonizing the amounts of the constituents with the various relevant rate constants so as to give an eventual steady state.

When different cell constituents and associated enzyme activities wax and wane from multiple causes and with independent frequencies, there will exist a certain probability that different functions (amounts or activities) will pass within a given short interval of time through a minimum value together. This probability is independent of the previous history of the cell, just as the chance in a radioactive atom of that conjunction of events leading to disintegration is independent of time.

Suppose now that when a given number of cell functions simultaneously drop to a minimum, viability is irretrievably impaired, that restitution becomes impossible and that death ensues. Then the chance of death in a given interval of time is more or less independent of the previous history of the cell. The condition for the approximate applicability of the exponential decay law is then fulfilled.

This point of view, which stresses the dynamic nature of the economy even in the non-proliferating cell, seems to offer a reasonable picture of the slow decline of a population. It does not ignore the influence of progressive decay of the whole and does not give the exponential law a higher status than that of an approximation. What part phenomena of the kind discussed might play in the rapid killing of a cell population by high energy radiation is difficult to say, but the picture presented of the cell dynamics in the non-proliferating state generally seems very well worth consideration.

DISCUSSION

M. R. Droop. Does Professor Hinshelwood consider that there is any profit in the analogy between ageing in the metazoan organism and the bacterial culture as opposed to the bacterial cell?

Sir Cyril Hinshelwood. I am not in a position to say anything very definite about this, but I do think that the relation of individual and community is one which could be profitably explored from a general point of view. Some groups are individuals from one point of view and communities from another. Certainly there are most interesting phenomena to be observed in the ageing of bacterial colonies, which themselves possess a certain unity and individuality. In liquid cultures the bacteria are more nearly independent units but by no means completely so, and their mutual interactions are significant in several ways.

O. V. S. Heath. By appropriate modification of the solid medium on which yeasts are grown, the cells can be made not only to grow abnormally large but to fail to separate after division, thus forming a sort of tissue. This suggests that the line between colonies of unicellular organisms and multicellular organisms is not a hard and fast one.

AGEING IN HIGHER PLANTS

By

O. V. S. HEATH

Research Institute of Plant Physiology, Imperial College London, S.W.7

THE mode of growth of higher plants differs so fundamentally from that of animals that the problems of ageing may be expected to differ in like degree. A perennial plant, such as a dicotyledonous or coniferous tree, has all the working parts continually renewed, in marked contrast to a vertebrate animal where for instance the same nerve cells have to function throughout the life of the individual. Such a tree consists essentially of a living layer surrounding a double cone of dead wood and surrounded in turn by a skin of dead bark. The aerial half of the double cone bears at the top a succession of more or less ephemeral living organs, the leaves (and from time to time, flowers); the subterranean half bears at the bottom a succession of even more ephemeral living organs, the root hairs. The branching which occurs in both halves makes possible increased areas both of leaf surface and absorbing root surface (which is confined to the youngest part of the roots) but does not affect the validity of this simplified general picture. The growth in length of each half of the double cone results from cell division in a terminal meristem (or embryonic zone) and from subsequent extension of the new cells so formed; the growth in girth arises mainly from new cells formed by a continuous meristematic layer, the cambium, which develops a short distance behind the terminal meristem and completely surrounds the shoot or root as the case may be. This cambial layer gives rise to new tissues both towards the centre and towards the exterior, thus forming new wood and bast respectively. The terminal meristem of the shoot is responsible not only for the growth in length but produces a succession of primordia which develop into new leaves; in the axils of these, buds are formed which may expand to give new branches. At times the terminal meristem, generally that of a special branch of limited growth, may become modified to produce an inflorescence.

It follows that all the cells living at any given time will a few years later have been added to the dead wood or bark, or have become detached from the tree with the dead leaves or flowers or with the (living) fruits. (A partial exception is provided in some cases by wood parenchyma cells, which in *Pinus radiata* for example may survive for as long as seventeen years (Harris, 1954), but these do not seem to play any active part in the life of the tree.) Thus the growth in some respects resembles reproduction, for the tree will have become an entirely new individual in the sense that a plant from a cutting or a tissue culture transplant is a new individual. The tree is in fact not unlike a coral colony though a much more highly organized " symbiotic community " (J. E. Baker) and apart from accident, disease and some mechanical and nutritional considerations there seems no reason why it should not live indefinitely. The giant redwood trees of California (*Sequoia gigantea*) show that this is not far from the truth, for specimens are estimated to have lived for some 3,000 years, attaining a height of over 300 feet and a diameter at breast height of 24 feet. Specimens of swamp cypress (*Taxodium mexicanum*) are estimated to be 4,000 years old.

It would seem that the meristems should be ageless as they are renewed continually. Nevertheless they are found to show systematic changes with time and although some of these are undoubtedly caused by external factors others seem to be due to internal factors only. The periodical changes in the activity of the *cambium*, which give rise to " spring wood " and " summer wood ", thus forming the well known annual rings in timber, seem to be intimately connected with bud expansion and dormancy (Wareing, 1951). Some tropical timbers grown in non-seasonal climates show no annual rings, but even so there is normally a pause in cambial activity with the shedding of leaves and a renewal of activity with the expansion of new leaves (Simon, 1914, Coster, 1927–8 both cited by Richards, P. W., 1952). Thus cambial activity appears to be geared to that of the terminal meristems. Leaf-fall and leaf expansion are in many species controlled by the environment (especially day length), but in others appear to follow inherent rhythms. The *terminal meristems* themselves seem to be subject to inherent changes with time, not only in the periodical production in perennials of new leaves and of flowers but also, especially perhaps in annuals and biennials, in the kinds of leaves produced. In the latter types of plant the change over from purely vegetative growth to reproduction is associated with profound changes in the terminal meristems and is followed, after fruiting, by death; even in perennials there is normally, with the first onset of flowering, a marked check to vegetative growth and to the activity of those terminal meristems which remain vegetative.

The main problems of ageing in higher plants are thus, in order of increasing importance:

(*a*) The ageing and shedding of individual organs.
This is intimately bound up with
(*b*) Inherent changes in rate of leaf production.
(*c*) Inherent changes in kinds of leaf produced.
(*d*) The change from purely vegetative growth to reproduction.

Discussion of the first two of these problems is omitted owing to lack of space.

Sequential changes in the kinds of leaf produced by the meristems

It is well known that the early leaves of many plants show " youth forms ", often of simpler shape or with better developed laminae than those formed later. This may apply to plants grown from cuttings as well as to those grown from seed. Thus cuttings of the hare-bell (*Campanula rotundifolia*) revert to the rounded leaf form for the first few new leaves, the later ones being linear. Early leaves produced by cuttings of the garden geranium (*Pelargonium zonale*) are much larger, thinner and less xerophytic than those produced later; it has not been found possible to prevent this change by cultural means and it may be connected with the initiation of flowers (see below). In many cases a reversion to the " youth form " can, however, be produced by manipulation of external factors. Thus the hare-bell can be made to produce rounded leaves by low light intensity and the shrubby veronicas of New Zealand will produce leaf blades, which they lack in the mature form, in response to high humidity. The retention of the broad " youth form " of leaf by eucalypts grown in this country may be due to the same factors of low light intensity and high humidity. The fact remains that external factors cannot, apparently, *prevent* the appearance of " youth forms " in early formed leaves and thus a change in internal factors is implied. Ashby and Wangermann (1950) conclude from results of a sowing-date experiment that the gradients of cell size and cell number in successive leaves of *Ipomoea purpurea* are not wholly due to external factors, but are " symptomatic of some process of ageing in the apical meristem ". In ferns there is evidence that small size of the apical meristem leads to the production of juvenile leaf forms (Allsopp, 1953).

It is less generally realized that the successive leaves produced at the terminal meristem may, whether or not they show striking differences in shape, have markedly different rates of metabolism. This has been shown for barley by Gregory, Richards and their co-workers. Assimilation rates and respirations rates of successive fully expanded leaves showed characteristic differences, which were modified by manurial deficiencies (Gregory and Richards, 1929); in the fully manured plants each rate for the last-formed (10th) leaf was about half of that for the 3rd leaf. In later work (Gregory and Sen, 1937, Richards, F. J., 1938) respiration rates and other things were determined at later stages for each leaf as well as at its time of full expansion. These show that not only does each successive leaf produced by the terminal meristem start with a different respiration rate, water content, protein content, sugar content, etc. (all of which are modified by manurial treatment), but that the time drifts of these characteristics as the leaves age, differ from leaf to leaf and again are modified by manuring. Such results led Richards, F. J. (1934) to call attention to the dangers of treating a series of successive leaves present on a plant at any one time as if it constituted a simple age series. The variation from leaf to leaf may often be much greater than the change with age in any one leaf. The apparent " age drift " obtained from such a series of successive leaves is compounded of the effects of the positions of the leaves on the plant and of their ages; it bears no simple relation to the true time drifts for individual leaves nor to the mean time drift for all the leaves. This warning is still frequently ignored.

The change from purely vegetative growth to reproduction

Some aspects of this change will be considered in rather more detail and since it greatly affects growth rates a preliminary consideration of growth relations may be useful.

In higher plants the rate of growth in dry weight at an instant will be the net gain represented by the difference of the rate of assimilation (X) and the rate of respiration (Y); it will be completely determined by:

1. The assimilation rate (per unit leaf area) $= x = \dfrac{X}{A}$
2. The respiration rate (per g. of total dry weight) $= y = \dfrac{Y}{B}$
3. The ratio of leaf surface to total dry weight $= \dfrac{A}{B}$
4. The total dry weight itself (B) which together with *3* specifies the leaf area A.

If the first three were to remain constant over a period, then the assimilation rate would also be constant if expressed per unit of total dry weight (in view of the constancy of *3*); hence the net gain per unit of total dry weight would be constant, i.e., the rate of increase in dry weight would be proportional to *4*, the dry weight already achieved. The *relative growth rate* would thus be constant and growth would follow an exponential curve (Blackman, 1919). Since plant growth is largely a process of cell multiplication, this is the curve to which it will tend *if* all the cells, or a constant proportion of them, continue to divide and *if* they do so at a constant rate, i.e., if growth continues unrestricted by internal or external factors. It seems most improbable that either of these two provisos would be fulfilled for any long period. Further, in view of the differences in assimilation rates and respiration rates of successive leaves already noted, and of the mechanical necessity for a greater proportion of supporting tissue as the total leaf surface enlarges, it seems equally improbable that the values of *1*, *2* and *3* above would remain constant. Nevertheless it is found in many cases that the relative growth rate *is* nearly constant, falling very slightly until flowering ensues when a sharp drop occurs. This was noted by Blackman in the paper cited and he called the relative growth rate in terms of total dry weight the " efficiency index " since it measures the efficiency with which the plant uses its " capital " as represented by the total dry weight.

Data for the cotton plant (Heath, 1937*a*) presented in Figs. 1 and 2 show approximately exponential growth until flowering ensues, both for total dry weight and (after the first 3 weeks) for height. The more rapid initial growth in height may be attributed to extension

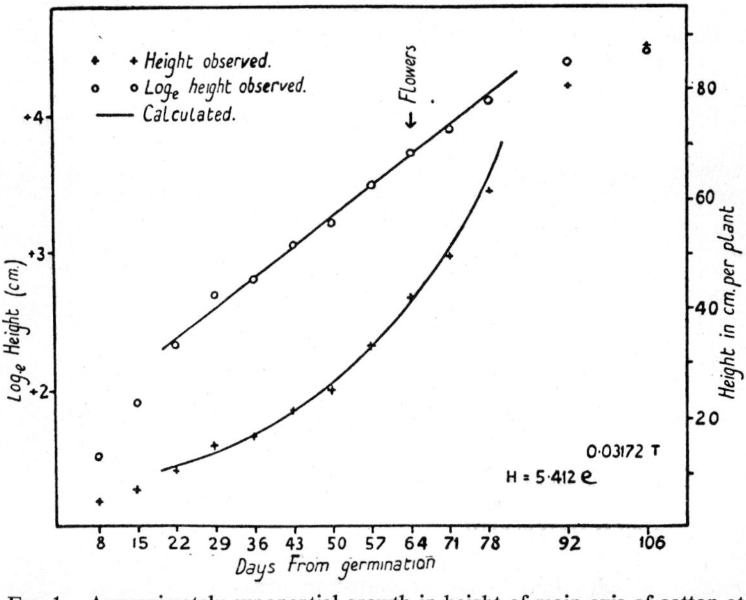

FIG. 1.—Approximately exponential growth in height of main axis of cotton at Barberton, South Africa, season 1933–4. (From *Annals of Botany*, N.S., **1**, 1937, p. 517.)

growth of the hypocotyl making use of reserves from the cotyledons. It may be noted that the relative growth rate in terms of height (a linear measure) is one-third of that in terms of weight (a cubic measure). The changes with age, up to first flowering, of other derived rates give a clue to the almost constant efficiency index (Heath 1937*b*): net assimilation rates, calculated by Gregory's (1926) method but on a basis of leaf weight instead of leaf area, showed no significant upward or downward trend. This measure is the resultant of Nos. *1* and *2* above (assimilation and respiration rates) and its apparent constancy may have been partly due to a simultaneous fall or other change in both rates for successive leaves. On the other hand relative growth rate in terms of total dry weight (efficiency index) did in fact exhibit a small but significant downward trend, as shown by the quadratic regression of log {total dry weight} on time (Table 2). Since net assimilation rate showed no downward trend, this fall in

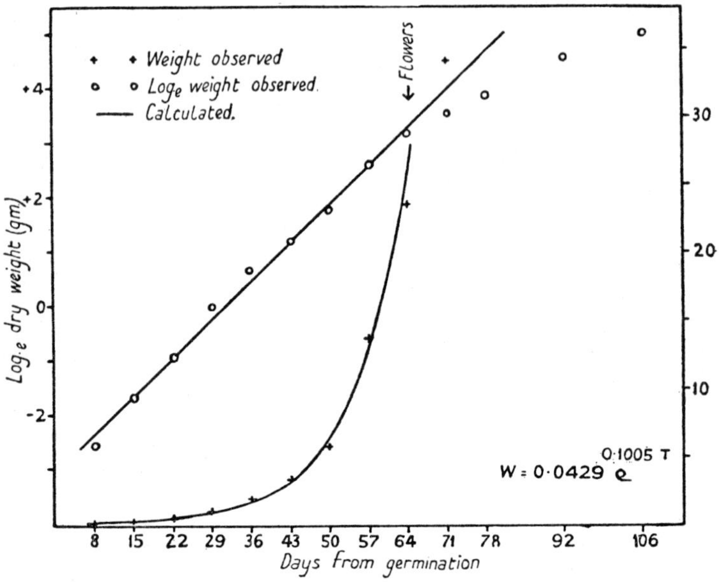

FIG. 2.—Approximately exponential growth in total dry weight of cotton at Barberton, South Africa, season 1933–4. (From *Annals of Botany*, N.S. **1,** 1937, p. 517.)

efficiency index had to be attributed to a marked decrease which occurred in the ratio of leaf to whole plant (Table 1; data for two seasons are included).

TABLE 1.—(from Heath 1937*b*)

Leaves as percentage of total dry weight of shoot

Season	*Initial sample*	*At flowering*
1933–4	77·0	54·7
1934–5	74·9	56·4

(It should be noted that this decrease would tend to cause a downward trend in net assimilation rate, for with the passage of time a greater proportion of the plant would contribute to respiration only and not to both assimilation and respiration). This decrease in the proportion of leaf resulted in a simultaneous fall in relative leaf growth rate. The fall in efficiency index was less rapid than this last however (Table 2), because the increasing proportion of assimilates going to parts of the plant other than leaf was included in the total dry weight but not in the leaf dry weight. Thus a very marked decrease in the proportion of leaf causes a much less striking decrease in efficiency index.

TABLE 2.—(from Heath 1937*b*)

Mean quadratic regression coefficients (14-day) for logarithms of total dry weight and of leaf dry weight

Season	*Total weight*	*Leaf weight*	*P**
1933–4	—0·0587	—0·0751	0·001
1934–5	—0·0464	—0·0545	0·1—0·2

*P=probability of the difference between the two measures being due to chance.

Up to the first flowering, then, the efficiency index may be expected to fall rather slowly with time; the drop which follows is altogether more dramatic. An interpretation in terms mainly of nitrogen nutrition has been proposed by Gregory for barley (Gregory, 1937; Goodall & Gregory 1947), and adopted by Crowther for cotton (Crowther, 1934). The latter case especially will be considered here. Crowther postulates a " self-regulating mechanism " essentially as follows:

In the early stages of growth there are few positions where buds can form and thus even a low external nitrogen supply suffices to give the maximal rate of development for the existing environmental conditions of temperature, etc. Owing to the exponential nature of the early growth the number of positions for growing points, where nitrogen is utilized, is rapidly multiplied. Thus at all external nitrogen levels the internal concentration falls, the rate of fall being determined by (1) the concentration in the soil solution and (2) the rate of development of the plant. This last is mainly controlled by other environmental factors (e.g. temperature) as long as the internal nitrogen supply is adequate, but ultimately this is no longer the case and nitrogen becomes severely " limiting ". Gregory has called this the stage of " internal starvation " and takes it to coincide with the stage of maximum leaf area, after which the loss of old leaves by senescence is more rapid than the formation of new ones. Crowther found that this occurred with a concentration of 2·5 per cent nitrogen (on a dry weight basis) in the green leaves, whatever the level of nitrogen manuring. He considered the plant to be a victim of its own morphological processes—the more nitrogen it was given the more meristems it formed and the larger the number of flower buds it produced. Boll (fruit) development followed with the resultant severe drain on nitrogen and also carbohydrates. Growth then ceased, and many of the flower buds and bolls were shed. Crowther

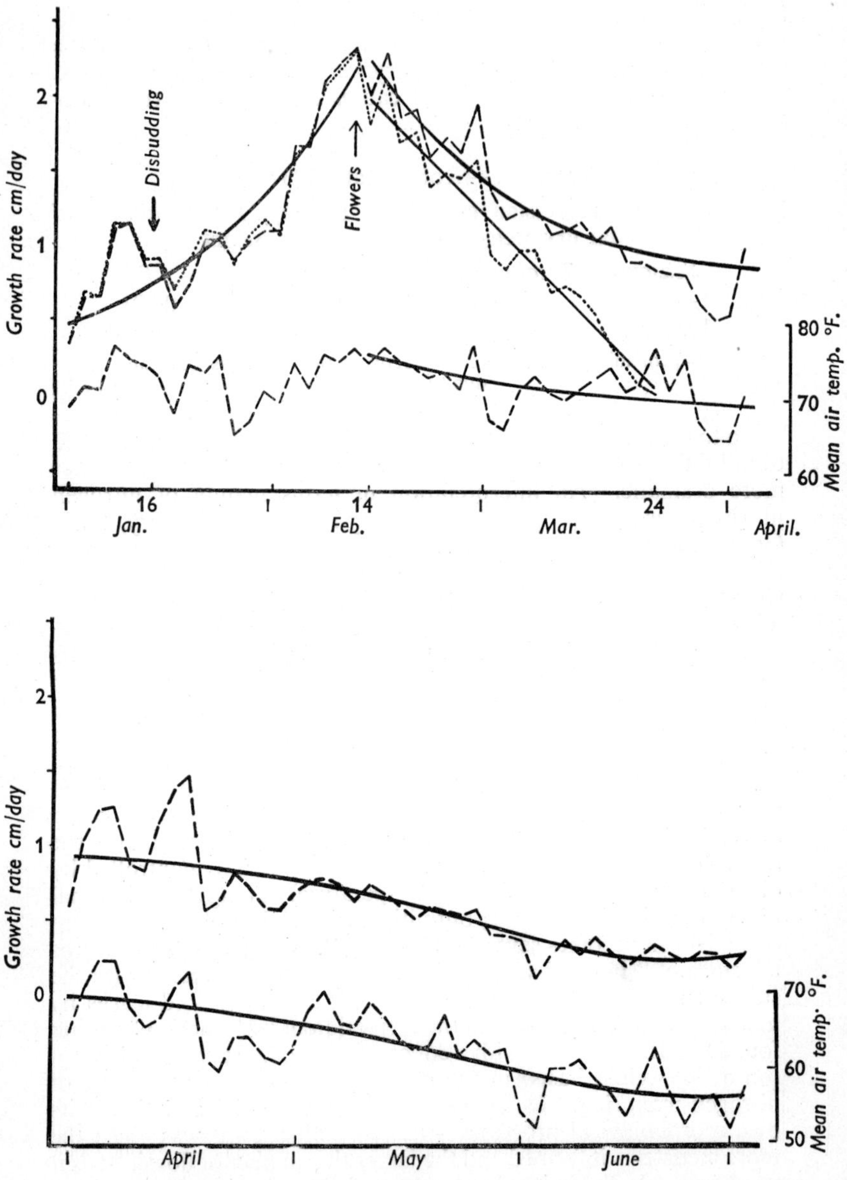

FIG. 3.—Absolute growth rates in height of main axis of " normal " and " disbudded " cotton plants; also mean air temperatures, Barberton, South Africa, season 1930–31.

found that simultaneously with the rapid transference of nitrogen and carbohydrate to the developing bolls, uptake of nitrogen from the soil dropped almost to zero, even with much soil nitrogen remaining. He attributed this to carbohydrate starvation of the root system, following boll development, preventing growth of the roots.

Crowther's data for distribution of nitrogen in the different parts of the cotton plant show clearly that much or most of the nitrogen for the bolls is withdrawn from the leaves. Nevertheless, the check to growth which occurs with the onset of flowering does not appear to be solely a nutritional effect. In Fig. 3 are presented hitherto unpublished data for the mean (absolute) growth rates in height of 45 normal cotton plants and also for 45 similar plants on the same plot from which all flower buds and developing monopodial lateral shoots were removed every alternate day, when all the plants were measured. In the disbudded plants, then, the nutritional strain associated with flowering should have been in the main prevented. Up to the day of first flower opening in the normal plants the same curve, calculated from an exponential curve, fits both sets of data almost equally well; the removal of buds for the greater part (28 days) of this period had no noticeable effect on the growth rate of the disbudded plants. Following first flower opening the growth rate of the normal plants fell almost linearly to zero in 40 days; the growth rate of the disbudded plants *fell from the same date*, scarcely less rapidly for the first three weeks, and from the time when the normal plants ceased growth (24.3.31) it followed a trend almost identical with that of mean air temperature. The curves shown for growth of the disbudded plants and mean air temperatures are fourth degree polynomials. The total correlation between the growth rates and corresponding mean air temperatures from 24.3.31 to 2.7.31. gives a value of r of $+0{\cdot}865$, or from 14.2.31. to 2.7.31. of $+0{\cdot}838$. It seems clear that for the disbudded plants, after flowering of the normal plants, temperature was the main factor controlling growth rather than nitrogen or other nutrients. The sharp falling away from the exponential type of growth at the time for first flower opening, even in the absence of any flowers, strongly suggests some inherent change in the terminal meristem, possibly hormonal in nature. Data for disbudded plants were collected in one season only and it would be desirable to see such an experiment repeated elsewhere; nevertheless they are at least suggestive.

It may be noted that in barley the stage of maximum leaf area, the fall in nitrogen uptake and the falling off from exponential growth coincide approximately with *initiation* of the inflorescence (Gregory, 1937, 1949) (not with filling of the ear), and again a purely nutritional explanation would seem inadequate to account for this fact. Possibly hormone production associated with such inflorescence initiation may check leaf growth and root growth. This is a different stage from that found for cotton, for first initiation of flowers occurred more than four weeks before first flower opening in the experiment just considered. However, since barley is a monocotyledon with a terminal inflorescence and cotton a dicotyledon with a system of sympodial lateral flowering shoots some other differences in organization may also be expected.

Finally, the " minimum leaf number ", found before the occurence of inflorescence initiation in certain plants with terminal inflorescences, may be briefly discussed. Klebs (1918) postulated that until a plant had reached a certain stage of development and was " ripe to flower " no combination of environmental conditions could make it flower; after this stage it would flower in response to appropriate external-conditions only. Clearly, if such a stage is rigidly defined it may at any time be disproved by the experimental exploration of new combinations of conditions, like any other null hypothesis. Thus Purvis (1934) and Purvis and Gregory (1937) found that in rye the number of leaves preceding the terminal inflorescence could be increased by environmental conditions up to about 25, but could not be reduced below seven. This was therefore accepted as morphologically defining the stage of " ripeness to flower ". Much later Gott, Gregory and Purvis (1955) found that this number could be reduced to five by the use of continuous illumination. As the rye embryo already has four leaf primordia in the ripe grain and another develops during sprouting, this raised the possibility that the plant was " ripe to flower " immediately after germination and as soon as the first leaf emerging was exposed to light. By prematurely harvesting grain, dwarf embryos were obtained with as few as two or three leaf primordia but the minimal number of leaves formed before flowering was still five. This was taken as evidence for an obligate vegetative stage, represented by the development of five leaves, which must precede flowering.

In the onion the minimum number of leaves and scales so far found before flowering is about 13 (Heath and Holdsworth, 1948) and this leaf number may perhaps represent an obligate minimum for " ripeness to flower ". There is, however, a marked effect of *size* of plant on inflorescence initiation, whether in plants prevented from bulbing by short days (Holdsworth and Heath, 1950) or plants which have bulbed (Heath and Holdsworth, 1948). The evidence suggests that initiation follows when the plant has attained a sufficient size, and the temperature is sufficiently low (a mean of about 12°C.), irrespective of *total* number of leaves and scales produced or of whether the plant has formed a bulb. It is most unusual for inflorescence initiation to occur in onion sets (small bulbs) formed from plants which had fewer than five *emerged* leaves and this may well be a matter of integrated leaf area. Experimental evidence was obtained for the production of an inflorescence-promoting hormone, produced in the emerged leaves, stored in the swollen leaf bases formed when bulbing occurred and transferred slowly to the growing point during winter storage. Small sets when ripe normally have no swollen leaf bases but only swollen scales; larger sets and bulbs such as normally initiate inflorescences have one or more swollen leaf bases according to size. It was found that large sets or even large bulbs could be prevented from initiating inflorescences by removal of the swollen leaf bases soon after harvest, leaving only swollen and unswollen scales and leaf initials none of which had ever had blades exposed to light. Such removal of swollen leaf bases was much less effective when carried out in December-January than when carried out in August directly after harvest, indicating the gradual transfer of something to the rest of the bulb and the growing point in the interval. Small bulbs, selected as of the same weight as the large bulbs immediately after the removal of leaf bases, gave more flowering than the latter, indicating an initially greater quantity of the inflorescence-forming substance in the swollen leaf bases than elsewhere in the bulb.

Thus in the onion the effect of age on inflorescence initiation would seem to operate mainly through size attained (possibly leaf area) and if there is an obligate minimal leaf number it is generally exceeded.

There is no lack of interesting examples of apparent age effects on the change from vegetative growth to reproduction, for instance the simultaneous flowering after many years of a plantation of Sisal (*Agave sislana*) followed by death of the plants, but it seems more fruitful to discuss experimental work when one is aware of its existence.

References

Allsopp, A. 1953. " Experimental and analytical studies of Pteridophytes XIX. Investigations on *Marsilea* 2. Induced reversion to juvenile stages." *Ann. Bot., Lond.,* **N.S. 17,** 37–55.

Ashby, E. and Wangermann, E. 1950. " Studies in the morphogenisis of leaves IV. Further observations on area, cell size and cell number of leaves of *Ipomoea* in relation to their position on the shoot." *New Phytol.,* **49,** 23–25.

Blackman, V. H. 1919. " The compound interest law and plant growth." *Ann. Bot. Lond.* **33,** 353–360.

Coster, C. 1927–28. " Zur Anatomie und Physiologie der Zuwachszonen und Jahresringbildung in den Tropen." *Ann. Jard. bot. Buitenz.,* **37,** 49–160; **38,** 1–114.

Crowther, F. 1934. " Studies in growth analysis of the cotton plant under irrigation in the Sudan I. The effects of different combinations of nitrogen applications and water-supply." *Ann. Bot., Lond.,* **48,** 877–913.

Gregory, F. G. 1926. " The effect of climatic conditions on the growth of barley." *Ann. Bot., Lond.,* **40,** 1–26.

Gregory, F. G. 1937. " Mineral nutrition of plants." *Ann. Rev. Biochem.,* **6,** 557–578.

Gregory, F. G. 1949. " The interaction of factors in determination of plant yield." Brit. Commonwealth Sci. Official Conf. *Plant and Animal Nutrition in Relation to Soil and Climatic Factors.* Melbourne.

Gregory, F. G. and Richards, F. J. 1929. " Physiological studies in plant nutrition I. The effect of manurial deficiency on the respiration and assimilation rate in barley." *Ann. Bot. Lond.*, **43**, 119–161.

Gregory, F. G. and Sen, P. K. 1937. " Physiological studies in plant nutrition VI. The relation of respiration rate to the carbohydrate and nitrogen metabolism of the barley leaf as determined by nitrogen and potassium deficiency." *Ann. Bot., Lond.*, **N.S. 1**, 521–562.

Goodall, D. W. and Gregory, F. G. 1947. *Chemical composition of plants as an index of their nutritional status.* Imp. Bur. of Hort. and Plantation Crops, Tech. Comm., 17.

Gott, M. B., Gregory, F. G. and Purvis, O. N. 1955. " Studies in the vernalization of cereals XIII. Photoperiodic control of stages in flowering between initiation and ear formation in vernalized and unvernalized Petkus winter rye." *Ann. Bot., Lond.*, **N.S. 19**, 87–126.

Harris, J. M. 1954. " Heartwood formation in *Pinus radiata* (D. Don.)." *New Phytol.*, **53**, 517–524.

Heath, O. V. S. 1937*a*. " The growth in height and weight of the cotton plant under field conditions." *Ann. Bot., Lond.*, **N.S. 1**, 515–520.

Heath, O. V. S. 1937*b*. " The effect of age on net assimilation and relative growth rates of the cotton plant." *Ann. Bot. Lond.*, **N.S. 1**, 565–566.

Heath, O. V. S. and Holdsworth, M. 1948. " Morphogenic factors as exemplified by the onion plant " in *Growth. Symp. Soc. exp. Biol.*, **2**, 326–350.

Holdsworth, M. and Heath, O. V. S. 1950. " Studies in the physiology of the onion plant IV. The influence of day length and temperature on the flowering of the onion plant." *J. exp. Bot.*, **1**, 353–375.

Klebs, G. 1918. " Uber die Blutenbildung von *Sempervivum*." *Flora,* **111/112,** 128–151.

Purvis, O. N. 1934. " An analysis of the influence of temperature during germination on the subsequent development of certain winter cereals and its relation to the effect of length of day." *Ann. Bot., Lond.*, **48**, 919–955.

Purvis, O. N. and Gregory, F. G. 1937. " Studies in the vernalization of cereals I. A comparative study of vernalization of winter rye by low temperature and by short days." *Ann. Bot., Lond.*, **N.S. 1**, 569–592.

Richards, F. J. 1934. " On the use of simultaneous observations on successive leaves for the study of physiological change in relation to leaf age." *Ann. Bot., Lond.*, **48**, 497–504.

Richards, F. J. 1938. " Physiological studies in plant nutrition VII. The relation of respiration rate to the carbohydrate and nitrogen metabolism of the barley leaf as determined by phosphorous and potassium supply." *Ann. Bot., Lond.*, **N.S. 2**, 491–534.

Richards, P. W. 1952. *The tropical rain forest.* Cambridge University Press.

Simon, S. V. 1914. " Studien über die Periodizität der Lebensprozesse der in dauernd feuchten Tropengebieten heimischen Bäume." *Jb. wiss. Bot.*, **54**, 71–187.

Wareing, P. F. 1951. " Growth studies in woody species IV. The initiation of cambial activity in ring-porous species." *Physiol. Plantarum*, **4**, 546–561.

Wareing, P. F. 1953. " Growth studies in woody species V. Photoperiodism in dormant buds of *Fagus sylvatica L.*" *Physiol. Plantarum*, **6**, 692–706.

DISCUSSION

Elizabeth Wangermann. There is more convincing evidence about inherent changes in meristems from plants which can be grown in a constant environment and which do not flower under experimental conditions. Thus, in the shoot meristems of *Lemna minor*, ageing changes definitely occur, as shown by a gradual, steady decrease in growth potential. Similar changes occur in isolated tomato root apices grown in culture medium, as shown recently by Professor Street. On the other hand, the apical meristem of *Salvinia natans* seems to go on growing indefinitely, without any loss in growth potential, under constant and favourable conditions. (Some of my plants have now been growing for five months at the same rate, and look like continuing indefinitely.)

It seems therefore that ageing changes at the meristem occur in some groups of plants but not in others, and each will have to be investigated separately.

F. G. Gregory. Dr. Heath pointed out that, in his experiment in which flower buds were removed, the relative growth rate remained the same in the two cases for a period of more than a month after flower initiation had begun. Such a result would not be unexpected if the internal level of nitrogen has to fall to some critical level before the growth of the apex is affected. This rate of decline of nitrogen would of course be affected by the removal of all flower buds, of which the nitrogen content is known to be high, and I suggest that this is one of the factors concerned.

The sudden fall in relative growth rate cannot be due to flower initiation, since that had already occurred and must therefore be due to some other event such as pollination.

The question of the mechanism of translocation of carbohydrate seems to me to be concerned. It may be that the direction of translocation is itself determined by internal factors and that in this case the determining event is associated with flowering. I suggest that once translocation is diverted from the roots this will continue even though the flower buds are removed. If anything of this kind were to happen one might perhaps account for the cessation of root growth and the problem would once again become one of nutrition.

O. V. S. Heath. Although the flower buds may have had a high percentage nitrogen content the total weight of buds was very small as they were removed at the earliest possible stage. It would be a curious coincidence if the amount of nitrogen so removed was equal to that taken up by the developing flowers on the normal plants, so that the critical internal nitrogen level was reached in both series on the same day, and furthermore that this should coincide with first flower opening.

On the question of direction of carbohydrate translocation the results obtained by Maskell and Mason of course suggest that this can be reversed by altering the positions of source and sink. Leonard has found for the tomato plant that root growth, which ceases with fruit setting, is resumed when the fruits are removed.

I agree that the contrast between the continued growth over many months of the disbudded plants and the rapid fall of the growth rate to zero in the case of the normal plants was doubtless due to the difference in nutrition brought about by the developing fruits of the latter. I cannot think, however, that the abrupt falling away from the exponential type of growth, which in both series coincided with first flowering of the normal plants, can be a matter of nitrogen (or other) nutrition alone and some sudden systematic change in the growing point seems to be implied. Possibly hormone production might then check root growth and so limit nitrogen uptake (thus making the mechanism once more one of nutrition as Professor Gregory has implied); even so, however, the change in the disbudded plants was induced, not by flower opening or pollination, but apparently by some systematic change which occurred concurrently with the flowering of the normal plants.

P. C. DeKock. Results obtained at the Macaulay Institute show that, in ageing leaves, the phosphorus/iron, potassium/calcium and citric acid/malic acid ratios decrease, as do also the respiration rate and the sensitivity of respiration to inhibitors, ageing leaves often showing a stimulation of respiration in their presence. This has been found for successive leaves on a shoot, as well as for the same leaf sampled at various times (tobacco, larch). There is considerable evidence to show that essentially similar changes in mineral composition, organic acids, respiratory rate and sensitivity to inhibitors, take place in all animal cells.

H. W. B. Barlow. May I present some of Mr. A. C. Mason's results relating to the effects of age and leaf position on the mineral composition of apple leaves. The information is presented in the form of a graph for each of the three elements phosphorus, calcium and iron. Successive leaves were sampled from stool-bed shoots at fortnightly intervals throughout the growing season. The gradual transition in the shape of the distribution curves may be noted and the essential difference between the curves obtained during the period of shoot elongation and those obtained when growth had ceased. The latter set of samples shows the effect of ageing alone on composition and it is seen to be slight compared with the effect of leaf position on the shoot.

Temperature affects the growth of coleoptile sections as follows: (*a*) the extension in 1 ppm. indoleacetic acid at 5, 10, 20, 30 and 35° C. follows an exponential course of growth at all temperatures, the final length achieved being least in the lowest temperature; (*b*) the curves for extension in pure water and indoleacetic acid at 10, 20 and 30° C. show that temperature and growth substance are to some extent interchangeable, the 20° C. water curve being very similar to that for 10° C. in indoleacetic acid.

When growth is rapid, extension is a good index of " age ", but at lower temperatures extension is slowed down more than " ageing ", so that these sections never attain the final length of those at higher temperatures.

Sections transferred from low to high temperatures never catch up those in high temperature all the time, and transfer of sections from water to indoleacetic acid cannot make up for the period of " ageing " in water; sections inhibited by, for example, iodoacetate, however, can catch up those in indoleacetic acid on transfer, suggesting that here ageing as well as extension has been arrested.

AGEING IN ANIMALS

By

ALEX COMFORT

Nuffield Research Fellow, Department of Zoology, University College London, W.C. 1

AGEING is the name given to the progressive decrease in vigour which occurs in some organisms with the passage of time, leading eventually to the death of the organism. A decline of this kind is probably not universal in multicellular animals, but it is common. For most purposes it is conveniently measured by determining from vital statistics the force of mortality, or the probability of dying, at different ages. In animals subject to ageing, the probability of death increases with age, so that these forms have characteristic maximum life-spans. In animals not subject to ageing, the probability of death would not increase, and the life span, though finite, would be indeterminate.

Ageing almost certainly occurs in all birds and mammals, and in those invertebrates whose capacity for cellular replacement is less than total (Comfort 1956a); it may not occur in invertebrates which continuously renew all their cells throughout life (e.g. sea anemones), and possibly not in those fish, reptiles and amphibians which grow continuously and have no fixed maximum size. Animals of sharply-fixed cell number which have no regenerative powers (rotifers, nematodes) tend to have an equally sharply-limited life span. One determinant of ageing may thus be the possession of cells which cannot be renewed by division and which deteriorate in the course of metabolic life. Mammalian age processes may well, however, involve the production of new cells differing in quality and in responses from those which they replace. Wear in irreplaceable structures such as teeth and wings, processes of accumulation, processes of depletion as in the non-feeding adult stages of some moths, and the effects of reproduction, all cause deterioration and death in various animals. Ageing in general must therefore be treated as a unity of effects rather than a unity of causes.

The rate of progress of age changes is a function not primarily of the passing of time but of developmental or metabolic activity (the " rate of living "). In many invertebrates, the life cycle can be prolonged by lowering the temperature, and even more by reducing the amount of available food (Ingle *et al.*, 1937). McCay (1952) has shown that if the immaturity of rats is prolonged by under-feeding, the animals can be kept fully active, but juvenile, for their normal lifespan and then re-started in growth, the time gained in marking time effectively doubling the length of life. The life of rats once sexually mature cannot be prolonged to anything like the same extent (McCay *et. al.*, 1941). These findings suggest that actuarial senescence is dissociable from total body metabolism just as growth and metabolism are dissociable in the tadpole. If this is so, one of the most important problems of age research is to determine the components of the "programme" of mammalian development which act as timekeepers in initiating the senile decline.

It might become easier to do this if we could obtain vital statistics for the greatest possible number of species, especially from the different types of vertebrate, so that age processes could be correlated with other physiological events. Such figures have been available, among vertebrates, only for man, a few rodents, and one or two agriculturally important animals (sheep, horses). There has been so far, for example, no complete life-table in existence for captive birds, whose longevity is known to be much greater in proportion than that of mammals similar in size and activity. In some cases there are large differences in rate of development and life-span between animals of closely related species which could profitably be investigated if better actuarial data were forthcoming.

One consequence of this lack of information is that we have no experimental mammal, intermediate in size and longevity between man and the small rodents, whose rate of ageing is accurately known. There are no published actuarial tables for rabbits or dogs, and only incomplete information for guinea pigs (Rogers, 1949). In consequence, many descriptions of physiological differences between "young" and "old" animals are in fact descriptions of differences between infants and youngish adults.

Apart from a limited number of species whose ages can be determined by inspection of annually-increasing structures such as scales or horns, it is evident that studies of ageing in long-lived vertebrates must be based chiefly on existing records. These include kennel-books and stud-books, notes kept by breeders, and the files of zoological gardens. Such material varies greatly in quality, and in most cases there are substantial losses to the record from sale, culling, or deliberate killing in the course of experiment. In spite of this, it has already been possible to obtain from these sources partial life tables for several breeds of dogs, and for a number of mammals and birds in zoological gardens. It is a common character of many of these mammalian populations that the curve of decline is almost arith-linear, and the distribution of age at death is therefore rectangular, a constant number of individuals dying in unit time (Comfort, 1956b and 1957). In dogs, figures for Pekingese, mastiffs and wolfhounds indicate a more rapid senescence in the large than in small breeds; the median and maximum ages of these pedigree breeds differ by less than 20 per cent from those recorded in dingoes and in wolves.

Wild populations of small birds and mammals, and of many insects, commonly behave as if they were ageless, because the standing mortality is so high as to be effectively age-independent, and few individuals live out their potential span. Haldane (1953) has shown that there is some evidence from ringing studies that the mortality in bird populations may actually fall with increasing age. There are, however, other animal populations, both vertebrate and invertebrate, where individuals regularly reach the specific age in the wild if they survive to adulthood—a constant age/mortality relationship cannot therefore be assumed unless evidence of its validity is available. Low adult mortality and sharp specific age under wild conditions has so far been found or inferred chiefly in the larger birds and mammals, but they also occur in several pluriennial molluscs (Allen, 1952 and Rapson, 1952), and may be more widespread than has so far been recognized.

The distribution of such patterns of age/mortality has an important bearing on the evolutionary interpretation which is placed upon senescence. The common factor of all the ageing processes is that they represent a progressive failure of adaptation and homoiostasis. Medawar (1952) has therefore suggested that biologically speaking the "cause" of ageing is a progressive decrease in the power of natural selection to ensure fitness at ages after the maximum contribution to the next generation of progeny has been made. Where few animals survive into late life, postponement of the expression of adverse genes may be equal in selective value to their complete abolition, and factors such as cessation of growth may be selected if they improve survival at the most fertile period of life, in spite of adverse effects in later periods. Weismann originally laid emphasis on the idea that senescence is an evolved adaptation. Neither interpretation yet fully explains the reasons for its frequent appearance throughout the animal kingdom, even in annual species where longer life and powers of overwintering in the fertile state might have been expected to be of selective advantage.

References

Allen, J. A. 1952. *J. marine biol. Ass.*, **31**, 515.
Comfort, A. 1956a. *The biology of senescence.* London: Routledge and Kegan Paul.
Comfort, A. 1956b. *Proc. zool. Soc. London.*, **127**, 27.
Comfort, A. 1957. *Proc. zool. Soc. Lond.*, in press.
Haldane, J. B. S. 1953. *J. Inst. Actu.*, **79**, 83.
Ingle, L., Wood, T. R. and Banta, A. M. 1937. *J. exp. Zool.*, **76**, 325.
McCay, C. M. 1952. In Lansing, A. *Problems of Ageing*. Baltimore: Williams and Wilkins.
McCay, C. M., Maynard, L. A., Sperling, G. and Osgood, H. S. 1941. *J. Nutrit.*, **21**, 45.
Medawar, P. B. 1952. *An unsolved problem of biology*. London: H. K. Lewis.
Rapson, A. M. 1952. *Austr. J. marine Freshw. Res.*, **3**, 170.
Rogers, J. B. 1949. *Anat. Rec.*, **103**, 498.

DISCUSSION

W. B. Yapp. Could Dr. Comfort tell us a little more of his reasons for making two statements which go rather beyond the evidence that he has presented to us? The first is that we have no life tables for birds, whereas there are in fact several for wild populations, so that he seems to imply the assumption that a life-table is only of value when it applies to captive animals. The second is that while it is probable that birds potentially suffer from old age some cold-blooded vertebrates may not; the life-tables of wild birds provide no evidence for the first part of this view, while the anecdotal evidence of long-lived reptiles and fish mean nothing more than that the potential maximum age is high.

A. Comfort. There are in fact life tables for birds in the wild, but it is not possible to study senescence in populations which do not live long enough to reach senescence. There is no life table for birds under protected conditions; and so far population work on birds such as the albatross, which probably reach senescence in the wild, has not been in progress long enough to be of significant use to us. I think it will probably prove to be correct that the majority of poikilotherms ultimately undergo senescent change. If we were to find animals in which the potential life-span were so long that it could never be exhausted by any significant number of animals in a population, enquiry about their potential senescence would be a little academic. Perhaps I may take indeterminacy of age to mean simply the behaviour of animals in which no increase in vulnerability with age has been detected however long they have lived.

J. L. Cloudsley-Thompson. Insects and other terrestrial arthropods kept at constant temperatures tend not to live so long as control animals in fluctuating environmental conditions. The subject has recently been reviewed (Cloudsley-Thompson, J. L., 1954, *Entomologist,* **86,** 183–9). Thus a steady decline in total activity during successive 24-hour periods occurs in cockroaches under constant conditions: but if two cockroaches, even of different species, are kept together this depression does not set in so rapidly (Cloudsley-Thompson, J. L., 1953, *Ann. Mag. Nat. Hist.* (12), **6,** 705–12.

R. L. Worrall. Since old people, generally speaking, have less protein than when they were young, there must be an overall excess of protein combustion over protein intake, for the period of adult life as a whole. This long-term net loss of body protein is too gradual and intermittent to be detected by short term nitrogen balance investigations. If a man contains 10 kilogrammes at the age of eighty, his average daily net loss of nitrogen is only 36 milligrammes, a figure well within the range of experimental error in nitrogen balance studies.

It is possible, however, to estimate the tendency of human adults towards protein deficiency, by measuring the percentage of total basal energy expenditure derived from protein combustion. I have published a convenient method of measuring this percentage in the Proceedings of the Nutrition Society for 1954.

By nutritional and other means, the long term net loss of body protein in human adults can be minimized, so that a serious degree of tissue atrophy is postponed, and active life is thus prolonged.

AGEING AND METABOLISM

By

FRANÇOIS BOURLIÈRE

Laboratoire de Physiologie, Faculté de Médecine, Paris

A LIVING organism being, as von Bertalanffy (1949) so aptly defined it, a graded structure of open systems which maintains itself in accordance with its inherent principles while its constituent elements are constantly changing, one may ask if the rate of renewal of these elements has some influence on the life-span.

It is now well known that the rate at which living beings expend the energy at their disposal varies throughout the life cycle and with various environmental conditions. This is especially apparent in poikilothermic invertebrates and vertebrates, which lack the efficient thermoregulating and other homoiostatic mechanisms which keep the milieu intérieur of homoiothermic vertebrates so remarkably constant and almost insensible to the usual variations of their environment. In such lower animals, the conditions which are theoretically able to modify the energy-turnover are numerous. The first to be considered—and the first which has been studied experimentally—is the environmental temperature. But many other factors are to be taken into account. Seasonal changes of activity and food consumption, hibernation, aestivation and reproductive activities are all able to influence the rates of vital processes.

Such a susceptibility to the variations of the environmental conditions is nevertheless not restricted to the " lower " animals. Temperature variations for instance may have an obvious influence on the metabolism of some birds (swifts, nightjars and humming-birds) and mammals (most marsupials, edentates, bats and even some Insectivora and lower Primates) and the ageing processes of such abnormal forms are well worth investigating from the standpoint of comparative gerontology.

It is the purpose of this paper to review briefly the evidence at hand on the correlation which seems to exist in the animal kingdom between the rates of energy-turnover and ageing.

Invertebrates

In spite of the scarcity of information on ageing processes in invertebrates a few well established facts seem to indicate that there is among them a definite and inverse relation between metabolic rates and speed of growth, and duration of life.

MacArthur and Baillie (1929*a*) used parthenogenetically produced males and females of one pure line of *Daphnia magna* and kept them at various temperatures (8, 10, 18 and 28°C.) the other conditions remaining constant. In such experiments the mean length of life of these cladocerans was strongly influenced by temperature. The correspondence between mean life-span and temperature is illustrated by the following figures: 25·586±0·28 days at 28°C., 41·675±0·35 days at 18°C., 87·8±1·98 days at 10°C. and 108·18±2·43 days at 8°C. The temperature coefficient for longevity averaged 2·12, which closely approximates that for most chemical reactions, but the Q_{10} values were higher (2·59) in the lower ranges and lower (1·64) in the higher ranges of temperature. It is to be noted also that the duration of life in males responded more sensitively and more extremely to temperature alterations than did that of females.

At the same time MacArthur and Baillie (1929*b*) found that elevation of temperature from 8 to 18°C. increased the heart rate and the susceptibility to potassium cyanide by a little over two and a half times, and this acceleration of metabolic rates was associated with a very nearly proportionate decrease in duration of life. It was also noted that the duration of life of females exceeded that of males, corresponding to a sexual difference in metabolic rates.

These results are confirmed by the findings of Terao and Tanaka (1930) on another cladoceran *Moina macrocopa.* The maximum duration of life of this species takes place at about 15°C. (14·36±0·38 days), and gradually decreases at higher temperatures (9·28±0·18 days at 21°C.; 6·51±1·67 days at 27°C.; 4·77±0·06 days at 33°C.). It is interesting to note that below 15°C. the life-span does not increase any more.

Alpatov and Pearl (1929) have investigated the effect of the temperature during development and imaginal life upon the duration of life of *Drosophila melanogaster.* Their results show that the effect of the temperature during embryonic, larval and pupal development upon the duration of imaginal life is as striking in insects as in crustaceans. A high temperature (28°C.) during development shortens the duration of the subsequent imaginal life at all temperatures and in both sexes, with but one exception. For instance females reared at 28°C. had a mean duration of life of 28·52±0·34 days, while the ones reared at 18°C. lived 70·61±0·98 days. The effect of temperature during imaginal life upon the duration of life of the imago itself is quite as striking; as the environmental temperature is higher flies live, on the average, a shorter time. This last finding confirms the results of earlier experiments by Loeb and Northrop (1917).

Another experimental way to influence the metabolic rate of invertebrates is by food restriction. Such a method has been widely used by experimenters and has given very consistent results. Using the two species of cladocerans already mentioned, Ingle (1933) and Ingle, Wood and Banta (1937) have found for instance that *Daphnia* which are starved live on the average about 40 per cent longer than those well fed, while their metabolic rate, as indicated by the heart beat frequency, is significantly lower than in normally fed individuals.

By alternating the feeding days with days of complete inanition Kopec (1924) succeeded in prolonging the larval life of the caterpillars of *Lymantria dispar* for a time which varied from 16·2 to 89·5 per cent of the average duration of the larval stage of control individuals, the prolongation being in direct proportion to the intensity of starvation. The pupal stage of the "restricted" animals was on the other hand distinctly abbreviated and the duration of life of the imago (which never takes food) remained unchanged. It thus appears that the prolongation of life of this insect by intermittent starvation only refers to its developmental stage. Subsequent experiments by the same author (Kopec, 1928) moreover show that food restriction of adult *Drosophila melanogaster* never increases the imaginal life span.

Nevertheless such a condition cannot be generalized to all kinds of arthropods. Among ticks, for instance, adult starvation is quite able to increase the longevity of the animal. Bishopp and Smith (1938) have thus found that while the adults *Dermacentor variabilis* which have attached to hosts usually die within a few weeks, the unengorged adults may live up to more than two and a half years—provided that they are kept in a suitable (moist and not too hot) environment.

Among rotifers the life span of the adults seems likewise to be inversely proportional to metabolic rate. Encysted adults may survive as long as 59 years in diapause (Rahm 1923), while most individuals live only a few weeks—from 8 days to 5 months, according to Comfort's table (Comfort 1956).

Besides these experiments numerous field observations appear to confirm both the action of temperature and the effect of food restriction and reduced activity on the duration of life of invertebrates.

The length of the whole period of development of many butterflies varies extensively with the environmental temperature. A tropical period studied by Fountaine (1938) in Cambodia reached its imaginal stage in only 17 days, the egg stage lasting two days, the caterpillar stage eight days and the pupal stage seven days. On the contrary there is evidence, accordance to Klots (unpublished) that larvae of some arctic Satyrids hatch from the egg in July, pass the next winter partly grown, spend the next summer feeding, hibernate again, and then transform to the pupa only in their third summer.

Differences in food availability among different populations of the same species may cause similar discrepancies in growth rate and life span. Such appears to be the case in *Patella vulgata* populations studied by Fischer-Piette (1939) in Brittany. In the estuary of the Rance river where food is abundant, the growth is very rapid and the limpets reach a very big size, but die when two-and-a-half years old. On the other hand those living on exposed rocks along the coast, when water is less rich in organic material, have a very slow growth, reach a smaller size but live much longer (up to sixteen years).

Vertebrates

The prolonged life span of most vertebrates does not favour an experimental approach to the supposed correlation between the rates of energy-turnover and ageing. We must rely on available information patiently gathered by ecologists and field-naturalists. It is not usually possible with such data to attain standards of precision comparable to well controlled laboratory studies. But they nevertheless give us some insight on the way the environment can differently influence the life-span of various populations of the same species.

Fishes offer remarkable opportunities from that point of view. Large samples of populations living in definite environments can be easily collected and an accurate determination of their age-composition can be made in most cases. On the other hand some species have a rather wide geographical distribution and live in waters whose temperature and food-content varies greatly.

In such conditions striking differences in growth rate, tempo of life and longevity can be found. The comparison of the results of Brown (1943) and Miller (1946) on two different populations of the North-American Grayling (*Thymallus signifer*) may exemplify this point. The Michigan subspecies (*Thymallus signifer tricolor*) living in Ford Lake has a very rapid growth and matures early, but has a short life. The oldest individual seen by Brown was in its sixth year of life. On the contrary the arctic subspecies found in Great Bear Lake (*Thymallus signifer signifer*) does not reach its sexual maturity before its fifth summer; it attains a greater ultimate size, and also a much greater longevity. Some individuals caught by Miller were in their twelfth summer. The length of life of the arctic subspecies is therefore about three times that of the southern one.

Similar instances could certainly be found easily in ichthyological literature. Sticklebacks (*Gasterosteus aculeatus*) do not live longer than 14 to 18 months in France, but require several years to reach their maturity in more northern latitudes. Pilchards (*Clupea pilchardus*) have a slower growth and, according to Flower (1935) live longer in the English Channel than off Saint Jean de Luz and Vigo.

Age determination in wild amphibians and reptiles being less reliable than in fishes, and in most cases quite impossible, the data on these two classes of vertebrates are scanty. Oliver (1955) has nevertheless given us recently some very interesting information on north American species.

The duration of the larval period of some widely distributed toads varies greatly with latitude. In North America the bullfrog (*Rana catesbeiana*) undergoes metamorphosis at the end of the first winter in Louisiana, while in the extreme northern part of its range (Nova Scotia) this species may spend three winters as a tadpole.

We have no data on the longevity of the adults in those cases, but we may infer from what is known that the duration of the whole life-cycle in the two cases is altogether different.

Among lizards striking latitudinal differences in growth-pattern and life-span are also known. In Florida, for instance, 94 per cent of the Southern Fence Lizards (*Sceloporus undulatus undulatus*) were found to live less than one year and none lived in their natural environment for two years. This warm climate permits the lizards to be active every month of the year with only brief irregular periods of inactivity resulting from low temperatures. In these conditions they grow throughout their short life-span. On the other hand, the northern race of the same species (*Sceloporus undulatus hyacinthinus*) usually lives, in Maryland, for more than four years, probably reaching a maximum age of eight years. Under this colder climate the Fence Lizard is active for only seven months of the year and its growth appears to stop after about the fifth or sixth year. It is therefore necessary to infer very cautiously differences between species with respect to determinate growth. In this case growth appears to be determinate in one part of the animal's range and indeterminate in another.

Similar data are lacking for snakes and turtles, most unfortunately since these are the longest lived of all vertebrates.

Seasonal variations of environmental temperature do not affect greatly the thermal constancy of the milieu intérieur in most warm-blooded vertebrates. The temperature factor has less effect on the energy metabolism of these animals than, for instance, food-restriction during the first stages of growth. Nevertheless one must not forget that the efficiency of thermoregulating mechanisms varies greatly in the various groups of birds and mammals. Some regularly enter periods of hibernation or of aestivation, during which their internal temperature may fall to a point close to that of the environment. In such states the metabolic rate is greatly reduced, often during many months a year. Others, like bats, have normally a very low metabolic rate, almost at the hibernating level, during about twenty hours a day, their oxygen consumption increasing only during flight. Such a periodical reduction of the energy metabolism, so well studied by Pearson (1947), is not at all restricted to the growth period but continues throughout life; it is significant to find that these mammals which have a poor thermoregulation and a normally low rate of metabolism are precisely those whose life-span is far longer than that of other mammals of similar size. In 1947 I reported the case of a *Rhinolophus hipposideros* ringed as adult in October 1938 and still living in fine condition in May 1946 (Bourlière 1947). More recently Dorst (1954) reported cases of marked *Rhinolophus ferrum-equinum* and *Miniopterus schreibersi* having reached respectively 15 and 14 years in wild conditions, potential longevities of 8 to 12 years being very frequent in these bats. It thus appears that in these mammals a reduced energy-metabolism and a low fecundity is definitely associated with a long life-span. Inversely, shrews, with their very high metabolic rate and fecundity are among the shortest lived mammals.

A very similar state of affairs seems to exist in some birds like swifts, hummingbirds and nightjars.

Such are the facts which suggest at the present some kind of correlation between rates of energy-turnover and ageing. Since the work of Rubner (1908) and Pearl (1928) our knowledge has slowly progressed and the influence on ageing processes of such factors as the "rate of living", the "tempo of life" and even the growth rate may well find their common denominator in a more or less direct action on metabolic rate. Most organisms seem unable to transform more than a certain fixed quantity of energy during their specific life-time and the rate of this energy-turnover appears to determine the rate of ageing.

If such a correlation is confirmed by more extensive experiments and observations it will remain to understand its underlying biochemical and biophysical causes. Some recent experiments showing that enzyme molecules have a definite life-span already suggest new approaches to that highly complex problem.

References

Alpatov, W. W. and Pearl, R. 1929. *Amer. Nat.*, **63,** 37.
Bertalanffy, L. v. 1949. *Das biologische Weltbild*, Bern, Francke.
Bishopp, F. C. and Smith, C. N. 1938. *Circ. U.S. Dep. Agric.*, 478.
Bourlière, F. 1947. *Mammalia*, **11,** 111.
Brown, C. J. D. 1943. *J. Wildlife Mgmt.*, **7,** 353.
Comfort, A. 1956. *The biology of senescence.* London, Routledge.
Dorst, J. 1954. *Mammalia*, **18,** 231.
Fischer-Piette, E. 1939. *J. Conch.*, **83,** 303.
Flower, S. S. 1935. *Proc. zool. Soc. London*, 265.
Fountaine, M. E. 1938. *Entomologist*, **71,** 90.
Ingle, L. 1933. *Science*, **78,** 511.
Ingle, L., Wood, T. R. and Banta, A. M. 1937. *J. exper. Zool.*, **76,** 235.
Klots (unpublished data).
Kopec, S. 1924. *Biol. Bull. Wood's Hole*, **46,** 1.
Kopec, S. 1928. *J. exp. Biol.*, **5,** 204.

Loeb, J. and Northrop, J. H. 1917. *J. biol. Chem.*, **32**, 103.
MacArthur, J. W. and Baillie, W. H. T. 1929a. *J. exper. Zool.*, **53**, 221.
MacArthur, J. W. and Baillie, W. H. T. 1929b. *J. exper. Zool.*, **53**, 243.
Miller, R. B. 1946. *Copeia*, 227.
Oliver, J. A. 1955. *The natural history of North American Amphibians and Reptiles.* Princeton, Van Nostrand.
Pearl, R. 1928. *The rate of living.* New York, Knopf.
Pearson, O. P. 1947. *Ecology*, **28**, 127.
Rahm, P. G. 1923. *Z. allg. Physiol.*, **20**, 1.
Rubner, M. 1908. *Das Problem der Lebensdauer und seine Beziehung zu Wachstum und Ernährung.* Munich, Oldenburg.
Terao, A. and Tanaka, T. 1930. *J. Fish. Inst. Tokyo*, **25**, 67.

DISCUSSION

J. B. Hamilton. The studies of Dr. Clive McCay, as mentioned by Dr. Comfort, suggest that an irreversible form of ageing occurs upon maturation of rats, whether the animals have been starved previously or not. Somewhat similar findings have been reported in some invertebrates (Kopec, 1924) but not in all (Bishopp and Smith, 1938). This point deserves further study especially with reference to phenomena of a non-sexual nature that occur at maturation. We have found that rates of protoplasmic replication in man, as judged by nail growth, are highest at adolescence. (Hamilton *et al. J. Gerontology*, 1955). Butcher has shown that in the rat a cycle of hair growth is a concomitant of maturation as judged by initial opening of the vagina and occurs at the appropriate time even in ovariectomized females. Similarly Dr. Heath reported this morning that the growth rate of certain plants is known to reach a peak at flowering and to decline thereafter, even if the budding flowers are plucked. I suspect that sexual maturation is but one phase—not the controlling force—of a phenomenon which McCay's work suggests to be crucial.

J. L. Cloudsley-Thompson. R. F. Lawrence (1953, *The biology of the cryptic fauna of forests*, Cape Town: A. A. Balkema, 301–3) has pointed out that most cryptozoic animals have an unusually long life span and he gives a table of examples that include the following: Mollusca (*Helix*) 8 years, Onychophora (*Peripatopsis* 6 or 7 years, Isopoda 3–4 years, Chilopoda (*Lithobius*) 5–6 years, Diplopoda (*Tachypodoiulus*) 2–6½ years, scorpions (*Buthus*) 6–7 years, false-scorpions 3 years, and spiders (*Eurypelma*) 20 years. It is tempting to speculate that the comparatively long lives of cryptozoic animals may be related physiologically to the fact that their food chains are based on humus which has a low food value, so that the fauna tends to have a low metabolic rate, as well as being correlated with a leisurely mode of life in a sheltered environment.

R. L. Worrall. An animal's energy turnover is geared to a continual building up and breaking down of body protein. When there is a long term excess of protein breakdown over protein formation, as in the case of an adult mammal, then the greater the energy turnover, the greater will be the net loss of tissue protein, and the sooner will come the time when the animal, or one of its organs, lacks the minimum amount of protein required for further life. A net loss of protein in old animals has been demonstrated for a wide variety of tissues, ranging from the brains of bees to the muscles of men. The length of adult life, in man and other mammals, is thus a function of three variables:

1. The energy turnover.
2. The net loss of protein per unit of energy turnover.
3. The minimum amount of protein required for life.

With approximate numerical values for these variables in different animal species, the general factors determining the length of adult life can be handled mathematically.

The energy turnover of a growing mammal, on the other hand, is geared to an excess of protein formation over protein breakdown, and the period of growth is a function of the variables:

1. The energy turnover.
2. The net gain of body protein per unit of energy turnover.
3. The maximum amount of body protein allowed by negative physiological feedback.

Each of the above variables is itself a function of others, the definition of which is a major task of biology.

THE AGEING OF MAMMALIAN CELLS

By

GEOFFREY H. BOURNE

Department of Histology, London Hospital Medical College, London, E. 1

THE rates and the methods of ageing of mammalian cells appear to be different in various organs and tissues. Because of this fact Cowdry (1952) classified the cells of the body from the point of view of ageing as follows:—

1. Vegetative inter-mitotics. Cells whose existence begins with mitosis and ends with the next mitosis. Examples of this are the basal cells of the epidermis, primordial blood cells and the spermatogonia. The individual lives of these cells are so short that they show little evidence of age change.

2. Differentiating mitotics. Cells which come from 1. Each of these is slightly more specialized than its parent. For example the blood cells show a progressive accumulation of haemoglobin content as they advance from megaloblast through normoblast to erythroblast, and the cells of the various layers of the skin contain progressively more keratin as they approach the surface.

3. Reverting post-mitotics. These are specialized (differentiated) cells each modified for the efficient performance of a specific function. These cells may show senescence changes and die. However, following injury, even in quite old animals they may rejuvenate in the sense that they undergo a series of rapid mitoses, and the relatively undifferentiated cells that result can differentiate into normally functioning cells of that organ. The cells of the liver, kidney, thyroid gland and salivary glands come into this category.

4. Fixed post-mitotics. These are highly differentiated cells and do not appear to be capable of further division. Among these are included erythrocytes, striated muscle and nerve cells.

It appears that cells which undergo mitosis do not age; presumably the mitotic process has a rejuvenating effect. Synthetic processes of cells, particularly those concerned with the production of protein, appear to be associated with nucleic acids. It is well known that new deoxyribose nucleic acid is synthesized and incorporated into the nuclei of both daughter cells during the process of mitosis. This has been demonstrated by the use of radio-active phosphorus. (See the results of Pelc and Howard 1953.) It may be that once a cell reaches a certain size, growth (that is ability to synthesize more protein) comes to an end because the synthetic capacity of the existing DNA is outstripped. With duplication of DNA however, fresh growth can begin. Comfort (1954) has pointed out that a theory of mechanical ageing could be based upon the exhaustion of cell catalysts.

Although cells of various tissues age in different ways and rates certain changes seem to be common to many of them. Mitochondria appear to be fragmented in older cells, and it is much more difficult to demonstrate filamentous mitochondria as such in cells in which they are normally present in this condition, when the cells are taken from an old animal. Whether this is the result of increased fragility in response to mechanical handling of the tissue or to the fixatives used or whether they are in this state in the living cell cannot be stated with certainty. Furthermore, in old cells it is always more difficult to demonstrate mitochondria because there appears to be a gradual accumulation in the cytoplasm of substances giving a similar staining reaction—these substances are probably phospholipid in nature. Bateman and Papez (1951) have shown an accumulation of lipids in old nervous tissues. This lipoidal material appears to be extruded from the cell and becomes embedded in the walls of blood vessels where it may give rise to sclerotic patches.

In the basophil cells of the pituitary in ageing fowls Payne has described changes in the mitochondria. At first they enlarge, then become vesicular. Many of them then fuse to form a single large body and at this stage the cell usually dies, the " chondrosphere " being liberated into the intercellular spaces. In the cells of the germinativum in ageing skin there is a reduction in the number of mitochondria. Tissue culture cells, provided they are constantly transferred into a fresh medium (of which embryo juices appear to be an essential constituent), undergo continuous growth and mitotic division, e.g. ten years of culture involving nearly 2,000 generations of cells is not unusual. If sub-culture is not carried out, however, degenerative changes followed by death occur. Characteristic amongst these are mitochondrial changes. These normally filamentous organelles break up into rods and granules or swell into small vesicles and there appears to be a nett decrease of mitochondrial substance. It may be significant that similar changes occur in the mitochondria in scurvy.

There are striking changes in the morphology of the mitochondria in parts of the senescent kidney, pancreas and liver, but these appear in localized areas mainly in association with sclerotic blood vessels. Apart from these areas the mitochondria seem unaffected, at least morphologically. Mitochondrial changes in the cells of the salivary glands in old animals have been described by Kurtz and Andrew (1951).

These mitochondrial changes are of considerable significance. It is known that something like 80 per cent of the enzymes of the Krebs cycle, which are responsible for the aerobic metabolism of the cell and which supply the energy required for oxidative phosphorylation, are found in these organelles ; substantial alterations in their form and quantity must have a fundamental effect on the metabolic activities of the cell.

Fragmentation of the Golgi apparatus is also characteristic of degenerative tissue culture and aged tissue cells. For example, Sulkin and Kuntz (1952) have shown that in the ganglionic nerve cells of young dogs the Golgi material is composed of a loosely woven net-work and is either restricted to the peri-nuclear portion of the cytoplasm or else extended right through it. In senile dogs this net-work like structure cannot be seen. Many cells show very little material with the same staining reactions as the Golgi substance, but where it occurs it is in the form of discrete granules. It is likely that electron microscope studies will tell us a great deal more about the structural changes in both Golgi material and mitochondria in old age.

In degenerating tissue culture cells characteristic nuclear changes occur. The nucleus may divide without mitosis into two or more parts (Andrew (1955) claims to have seen evidence of amitotic division in the nuclei of the cells of the brain in old animals). Vesicles of nuclear sap may be squeezed out into the cytoplasm. In some cells the nuclear wall may break down and the whole nucleus may liquefy. In old nerve cells the nuclei may become pycnotic and show a decrease in nuclear histone which, according to some authors, suggests a diminished capacity on the part of the nucleus to synthesize nucleic acids. In addition to pycnotic changes the nucleus may show an internal brushwork arrangement attached to one side of the nuclear membrane, and the nucleoli may show a honey-combed rim. Pycnotic nuclei have been described in many of the organs of old animals by a number of workers. Meyers and Charipper (1956) describe nuclear vesicles in the cortical cells of the adrenals of ageing hamsters.

In the cells of the autonomic ganglia of old dogs there seems to be a loss of chromidial substance (Sulkin and Kuntz 1952) which takes the form of a peripheral chromatolysis, the nuclei become pycnotic and displaced and the cells become shrunken and pigmented, also there is a shift in the nucleo-cytoplasmic ratio in the direction of the nucleus. It is of interest that sensory ganglia appear to be much more resistant to age changes for they do not show the features characteristic of ageing autonomic ganglia.

Cell membranes appear to alter with old age; certainly their permeability alters and presumably this indicates some structural modification. Not very much work has been carried out on the permeability of mammalian cells—Lansing's work (1947) on permeability of old cells was carried out with *Spirogyra*. It is of interest that he found that the alteration in permeability was not in the same direction for all compounds, for example, old cells were more permeable to alcohol and less to urea. This may also be true of the cells of the gastric mucosa of old human beings, at least for alcohol. As long ago as 1917 Herzfeld and Klinger showed that there was an increase of protein at the surfaces of old cells and that this interfered with the normal free exchange across the membrane. Lansing (1942) claims that there is also an accumulation of calcium at the membranes of old cells, and he states that by treatment which, he says, removes the calcium, he has been able to prolong the lives of rotifers. Calcium is known to be increased in amount in a variety of old human tissues, e.g. brain, sclera, arteries and elastic tissue (Simms & Stolman 1935). Lansing *et al.* (1949) have shown, using radio-calcium, that in old livers this metal is more firmly bound than in the young liver. They state that there is some evidence that some, at least, of the intracellular calcium is associated with a ribonucleoprotein complex situated at or near the cell surface.

Zondek and Karp (1955) showed that there is an increase of iron up to 200 per cent in the cells of old rats. This excess of iron reached a maximum in the rat at about 1½ years of age and thereafter it remained constant for the rest of the life of the animal (3–4 years). This increased iron content was found only in the cells of epithelial organs, not, for example, in muscle cells. The authors discussed whether this increase of iron was of special significance or whether it was dependent on the destruction of blood during the lifetime of the animal. They claimed that this increased iron was not due to haemosiderosis, pointing out that it occurred uniformly in organs in which haemosiderosis does not usually occur. The low values for iron in young animals and the high values for old animals were found to be constant. If the deposition of iron was due to destruction of blood one would expect the deposition of iron to be continuous throughout life—whereas this is not so. The authors quote this evidence therefore as proof that the increase of cell iron up to 1½ years of age must have some functional significance. They point out the remarkable constancy of their results; that the increase of cell iron at various ages is so regular that it is possible to tell the age of an animal by an iron estimation. They believe that the extra iron accumulated by the cell probably forms a compound other than those already present in the cell. It is of interest that changes occur in the haemoglobin iron of old mammalian erythrocytes (Lemberg & Legge 1949). In fact the ageing of erythrocytes is said to be due fundamentally to exhaustion of the system which maintains haemoglobin iron in the ferrous condition. It might be thought that this was due to the absence of a nucleus from the erythrocyte, but the avian erythrocyte, which possesses a nucleus and which appears to be able to synthesize haemoglobin continuously while the cell is in circulation, has a shorter life than the mammalian erythrocyte.

One aspect of ageing of the cells of some organs appears to be the accumulation of pigment. Hodge (1894) and Connor (1928) described the accumulation of yellow pigments in the nerve cells of man. " Senility " pigment has also been described in the ganglion cells of rabbits (Gatenby & Moussa 1950). These pigments appear first as caps around the nucleus and later spread throughout the cell. It is of interest that in vitamin E deficiency pigment occurs in many cells and some authors have thought that the so-called senility pigment may be due to this deficiency. However, there is evidence that the pigments of vitamin E deficiency are different from those of senility. The term " ceroid " has been loosely applied to senility and other pigments by a number of authors and it may be as well to clarify it. The word was first used by Lillie *et al.* (1942) for a type of pigment which is yellow to orange yellow in colour, which is not dissolved by fat solvents and yet retains a considerable sudanophilia even in paraffin sections. It was subsequently shown that this pigment was acid-fast (Endicott and Lillie, 1944). It also fluoresces a greenish yellow with the longer wave-lengths of the ultra-violet. Only pigment which has these properties is entitled therefore to be called " ceroid ".

Pigments of this type have been found in or around lipid deposits in atheromatous arteries of human subjects. It is thought by Hartroft (1953) that they form as a result of a reaction between the tissue lipid and the membranes of the red blood cells. Where small haemorrhages were present around lipid deposits in atheromatous arteries ceroid-like pigment was found invariably to be present. He obtained similar pigments from *in vitro* experiments with red blood cells and unsaturated fats. This has been confirmed by Casselman (1951). Pigment found in the cells of the autonomic ganglia of old dogs is often found to have considerable alkaline phosphatase activity.

One would expect that as cells age there would be a decrease in the various enzyme systems present in them. The falling off of oxygen consumption recorded by various authors (see Shock and Yiengst, 1955) suggests that this might be so. It may be of interest that Lindop (1956) found no difference in radio-iodine uptake in the thyroid gland of 246 rats varying in age from one day to 47 weeks old and in later experiments even in rats of much greater age. However Shock and Yiengst (1955) have shown a significant drop in basal heat production in old human beings.

Despite these findings, however, my own preliminary histochemical preparations of young and old animal tissues indicate a slightly increased succinic dehydrogenase activity in the latter suggesting that the energy of the Krebs cycle is being diverted into activities other than those of oxidative phosphorylation.

A further histochemical study has been made of the intensity of reaction of six types of phosphatases and of simple esterase in seventeen different organs in young and old rats. Preliminary observations indicate not a decrease in these hydrolytic enzymes in the tissues of old animals but rather an increase. This is most marked in the cerebrum and cerebellum (particularly in the Purkinje cells), in the brush border of the duodenum, and in the seminal vesicles; it was present but less marked in organs such as the liver and kidney. The substrates used for the phosphatases included glycerophosphate, which presumably demonstrates what is colloquially called " alkaline phosphatase " (which is, in fact, a non-specific phosphomonoesterase), oestrone and cortisone phosphates, pyridoxal phosphate, carbamyl phosphate and ethanolamine phosphate. The enzymes which dephosphorylate these latter compounds although having their optimum *pH* in the region of 8 or 9 appear to be quite distinct from the non-specific " alkaline phosphatase ". The histochemical preparations which are obtained from their use are quite different from those obtained with the latter.

So far as is known at present, oestrone and cortisone phosphates are not formed in the metabolic processes involved in the *in vivo* metabolism of these compounds, so that the increased dephosphorylation of them in older tissue may simply be an aspect of increase in general phosphorylytic activity in these tissues. On the other hand, of the remaining three, two (pyridoxal phosphate and carbamyl phosphate) are metabolically important substances, while the rôle of the third (ethanolamine phosphate), although it is widely distributed in the body, has not yet been elucidated; it may be associated with phospholipid metabolism. Pyridoxal phosphate forms the prosthetic group of several, and possibly all, amino acid decarboxylases, carbamyl phosphate is an important intermediary in urea synthesis in the body—it acts as a carbamyl donor in a variety of enzymatic systems. The fact that these substances are more easily dephosphorylated in old tissues may be of considerable significance, since it may indicate a shift in the balance between synthesis and hydrolysis of these physiologically very important substances towards the latter. It may be argued, probably quite justifiably, that histochemical preparations are artefacts and what goes on in them bears little relation to what happens *in vivo;* however, if we turn to the results obtained by the biochemists working with the characteristic artefacts in which they specialize, namely tissue homogenates, we find that they have recorded an increase in other types of hydrolytic enzymes in the tissues of older animals, namely β-glucuronidase (Byrbye and Kirk, 1956) and acid phosphatase (Zorzoli, 1955).

In this connection it is of interest that de Duve and his colleagues (1955) have described in cells of rat liver a series of bodies which appear to be microsomes and in which most of the hydrolytic enzymes appear to be concentrated. Among these enzymes were β-glucuronidase, acid phosphatase, cathepsin, ribonuclease and deoxyribonuclease, and the name given to these bodies was " lysosomes ". In a lecture at University College during 1956 de Duve described these bodies colloquially as " suicide bags " and claimed that their unrestricted activity after death was the cause of tissue autolysis. The increase in activity in these bodies in old tissues either by loss of inhibitory control or otherwise may help to explain the progressive inefficiency and degeneration of tissues as they age.

Cytolytic activity resulting from the activity of lysosomes may be responsible for the loss of neurones from the brain in older animals. This loss of neurones is one of the interesting aspects of the ageing central nervous system. Attention was drawn to this as long ago as 1894 by Hodge, who claimed a 25 per cent loss of Purkinje cells from the cerebellum of a man aged 92 compared with those of an adult man of 47. Harris (1927) said that in old apes there were only six Purkinje cells in senile tissues for every 41 in the young animal. Spiegel (1928) found a 40 per cent loss of Purkinje cells in senile guinea pigs and Andrew (1955) found in a woman of 80 years only 5 or 6 cells for every 20 in a young woman. Ellis (1920) listed the following numbers of Purkinje cells per given area of cerebellum in men of different ages: 42 years, 823; 65 years, 691; 79 years, 500; 94 years, 462; 100 years, 445.

Sulkin and Weatherford (1955) have shown that in autonomic ganglia in old dogs there is a decrease in vitamin C (as demonstrated by the acetic silver nitrate technique), changes in the Golgi apparatus, and accumulation of pigment and substances positive to the periodic Schiff reaction. They also showed that similar histochemical changes could be produced in guinea pigs on a vitamin C deficient diet. The dog is said to synthesize all the vitamin C necessary for its own requirements so in this case perhaps old age results in some decrease of the efficiency of the enzyme systems synthesizing this vitamin. It also raises the question as to whether partial deficiency of this vitamin over a very long period of time in human beings might not play a part in the initiation or acceleration of tissue changes usually attributed to the process of ageing.

Although it has been suggested that increase in hydrolytic enzyme activity in old age might be a fundamental factor in the production of the tissue degeneration of senescence we are still faced with the problem of the control of these processes. It is known that cells of a number of organs degenerate after removal of certain endocrines and it is also known that such an operation induces changes in enzyme activity and distribution in cells, e.g. Bourne and Malaty (1953) showed reduction of succinic dehydrogenase activity in kidney, liver and cardiac muscle following adrenalectomy and castration, and Brandes and Bourne (1955) showed loss of acid phosphatase from the Golgi region of the mouse prostate following castration, and its return to that site following implantation of testosterone.

Many papers in the literature have demonstrated the relations between morphological and biochemical changes in tissue cells and steroid and other hormones. In senescent humans and animals it has been shown that there are degenerative changes in the cells of the endocrine glands; Cooper (1925) many years ago demonstrated the changes in human endocrines with age. For example she found general increase of connective tissue in pituitary, adrenals and thyroid, in the pituitary she found an increase of basophil and chromophobe cells, in the thyroid loss of colloid or poor quality colloid and reduction in the size of cells. In the adrenal accumulation of what appears to be ceroid pigment and degenerative changes in the cells were found. A number of authors have studied the changes in these three endocrines in a variety of animals (see Blumenthal 1955) and degenerative changes in the reproductive glands are also well known and frequently recorded.

It may be, therefore, that tissue changes in age are due to progressive degeneration of the endocrine glands. Korenchevsky and his co-workers have shown that multi-hormone preparations can retard or reverse some but not all of the weight changes in organs due to age, and in some cases they aggravate the senescence changes.

Other workers have also studied the response of old tissue cells to hormones—Shock (1956) for example says that in tissues in old individuals which remain functional the response to steroid hormones judged by the N, K, P and Ca balances is the same as in the young subjects. Another investigation in this field was that of Masters (1953) who records the effects of sex hormone treatment of female patients who were over 60 years of age and who were all at least ten years past the menopause. These patients were followed until eleven of them died and autopsy material was obtained from the reproductive organs. The uterus and cervix were found to be increased in size and to be comparable in this respect with the normal for women between 20 and 40 years of age. The vagina was returned to a state identical in appearance with that seen in the fully functional vagina and reactivation of the cells of the breast was also obtained. Although microscopical section showed a reactivation of the blood vessels in the ovary the latter did not show a concomitant return to function.

Silberberg and Silberberg (1952) studied the relation between bone and cartilage age in relation to stimulation by hormones. They pointed out that the capacity for growth in epiphyseal cartilage declines with age and eventually ceases completely. On the other hand articular cartilage remains able to proliferate. Thus articular cartilage at any given age is in a younger state physiologically speaking than the epiphyseal cartilage. The latter has in fact undergone changes which are irreversible. Apart from this however, it seems that the majority of cells in the body although going through a process of ageing can still respond to appropriate stimulation.

This applies particularly to reverting postmitotic cells such as those of the liver, which have considerable regenerative power following injury or removal of part of the liver, even in very old animals.

It is possible that some of the age changes in nerve cells might be reversed with hormonal or other stimulation but no hormones will be able to bring back to the brain the cells which have already been lost by the ageing process.

If many of the ageing changes of tissue cells are due to failure of endocrine functions one might ask why it is that these latter organs degenerate. Although it is possible to say that they degenerate because of lack of secretion of trophic hormones from the pituitary gland there is still the query as to why the secretion of these trophic hormones decreases. Korenchevsky and Paris (1951) tried the effect of anterior pituitary hormones on the ageing female rat. They used an anterior pituitary extract which contained gonadotrophic hormones and also hormones somatotrophic for the preputial glands, the liver, kidney, spleen and heart, and for the other endocrines. The results obtained were a hypertrophy of the liver, heart and spleen, that is the hypotrophic process of old age has been reversed. When thyroid hormone was used in addition to the anterior pituitary hormones there was a beneficial effect on all the organs studied including the endocrines. However, although some ageing changes were reversed in some organs this combination of hormones did not appear capable of reversing the fundamental ageing processes which were taking place.

It appears therefore that some fundamental process is taking place in the cells of all organs of the body, including those of the endocrines, in ageing. Some of the changes in non-endocrine tissues may be due to lack of endocrine secretion and can be temporarily reversed with appropriate hormonal treatment but the ageing of the cells goes on. If, as mentioned earlier, it is due to progressive activity of the intra-cellular hydrolytic enzymes we must still ask " What is the cause of this increase ?" Concurrent with this enzymatic change there appears to be a decrease in adenosine tri-phosphate, a major source of energy for intra-cellular chemical reactions and, according to Lansing (1952), older tissues synthesize phosphoric esters with greater difficulty than the tissues of younger animals.

We are still a long way from understanding the mechanism which winds the life of the organism to a peak and then lets it slowly run down, but it is in the cell itself that we must seek the answer to this most perplexing problem.

References

Andrew, W. 1955. *J. Geront.*, **10**, 1.
Bateman, J. F. and Papez, J. W. 1951. *J. Geront.*, **6.**
Blumenthal, H. T. 1955. *J. Geront.*, **10**, 235.
Bourne, G. H. and Malaty, H. A. 1953. *J. Physiol.*, **122,** 178.
Brandes, D. and Bourne, G. H. 1954. *Brit. J. exp. Path.*, **35,** 577.

Byrbye, M. and Kirk, J. E. 1956. *J. Geront.*, **10,** 156.
Casselman, W. G. B. 1951. *J. exp. Med.*, **94,** 549.
Comfort, A. 1954. *Biol. Rev.*, **29,** 284.
Connor, C. L. 1928. *Amer. J. Path.*, **4,** 293.
Cooper, E. R. A. 1925. *The Histology of the More Important Endocrine Organs at Various Ages.* Oxford Medical Publications.
Cowdry, E. V. 1952. In Cowdry's *Problems of Ageing*, revised by Lansing. Williams & Wilkins.
De Duve, C. *et al.* 1955. *Biochem. J.*, **60,** 604.
Ellis, R. S. 1920. *J. comp. Neurol.*, **32,** 1.
Endicott, K. M. and Lillie, R. D. 1944. *Amer. J. Path.*, **20,** 149.
Gatenby, J. B. and Moussa, T. A. 1951. *J. Physiol.*, **114,** 253.
Harris, A. 1927. *Zool. Anz.*, **74,** 249.
Hartroft, W. S. 1953. *J. Geront.*, **8,** 158.
Hertzfeld, E. and Klinger, M. 1917. *Biochem. Z.*, **83,** 42.
Hodge, C. F. 1894. *J. Physiol.*, **17,** 129.
Korenchevsky, V. and Paris, S. K. 1952. *J. Path. Bact.*, **63,** 111.
Kurtz, S. M. and Andrew, W. 1951. *J. Geront.*, **6.**
Lansing, A. I. 1942. *J. exp. Zool.*, **91,** 195.
Lansing, A. I. 1947. *J. Geront.*, **2,** 228.
Lansing, A. I. *et al.* 1949. *Arch. Biochem.*, **20,** 125.
Lansing, A. I. 1952. In Cowdry's *Problems of Ageing*, revised by Lansing. Baltimore, Williams & Wilkins.
Lemberg, R. and Legge, J. W. 1949. *Hematin compounds and bile pigments.* New York, Interscience Publications.
Lindop, P. J. 1956. Lecture to British Society for Research on Ageing. London.
Lillie, R. D. *et al.* 1942. U.S. Public Health Report, **57,** 502.
Masters, W. H. 1953. *J. Geront.*, **8,** 33.
Meyers, M. W. and Charipper, H. A. 1956. *Anat. Rec.*, **124,** 1.
Payne, F. 1949. *J. Geront.*, **4,** 193.
Pelc, S. R. and Howard, A. 1953. *Exp. Cell. Res. Suppl.*, 11.
Shock, N. W. 1956. In *Hormones and the Ageing Process.* New York, Acad. Press.
Shock, N. W. and Yiengst, M. J. 1955. *J. Geront.*, **10,** 31.
Silberberg, M. and Silberberg, R. 1952. *J. Geront.*, **7,** 399.
Simms, H. S. and Stolman, A. 1935. *Science*, **86,** 269.
Spiegel, A. 1928. *Anat. Anz.*, **79,** 173.
Sulkin, N. M. and Kuntz, A. 1952. *J. Geront.*, **7,** 533.
Sulkin, N. M. and Weatherford. 1955. (Unpublished.)
Zondek, S. G. and Karp, J. 1955. *Biochem. J.*, **28,** 587.
Zorzoli, A. 1955. *J. Geront.*, **10,** 156.

DISCUSSION

G. C. Kennedy. Was the relatively slight difference between the young and old livers merely due to staining difficulty? Dr. Bourne commented that it was a less convincing slide than the rest. If the liver does show ageing changes, could not Dr. Comfort's point be investigated by doing a partial hepatectomy in an old rat and looking at the staining reactions of the new generation of regenerating cells?

G. H. Bourne. That is an interesting experiment which would be well worth doing.

N. W. Pirie. The importance of distinguishing between the age of a cell and the age of a tissue is well illustrated by the tobacco plant. The cells in most of the leaves of a plant a foot high are of comparable age because cell division finishes while the leaf is tiny and the development of successive leaves is mainly just cell expansion. During this expansion there is an increase in the proportion of the leaf phosphorus that is present as phospholipid and a diminution in the proportion present as ribo- or deoxyribo-nucleic acid (Holden, 1952, *Biochem. J.*, **51,** 433). The ribo-nucleoprotein becomes less easily extracted (Pirie, 1950, *Biochem. J.*, **47,** 164). These changes resemble those which take place in the liver as an animal ages but it may be that there too the individual cells may not differ so much in age as we tend to assume.

E. Digby. We are at a very early stage in the study of reversion, yet competent research workers are obtaining quite significant results. There was, for instance, the recent work by Dr. Pincus in America as recorded on Vol. XI of *Recent Progress in Hormone Research*. I would very much like Dr. Bourne to comment on it. Fourteen men in good health aged between 70 and 91 were treated by steroid replacement therapy. After four months they averaged 50 per cent increase in muscular strength, with heart action and pulse unaffected by the increased muscular energy. There was also a general sense of improved well-being. None of them showed untoward symptoms.

G. H. Bourne. I have no comment.

R. J. Ludford. Several references have been made during this conference to the significance of mitosis in the ageing of organisms. Dr. Comfort suggested that one determinant of ageing might be the occurrence of cells incapable of division, which deteriorate in the course of metabolic life, and Dr. Bourne recalled the reports in the literature of the loss of neurones from the central nervous system of ageing mammals. Dr. Bourne also stated in his paper: " it appears that cells which undergo mitosis do not age; presumably the mitotic process has a rejuvenating effect ".

The results of tissue culture work suggest that cells which retain the capacity to divide may continue to grow indefinitely, provided the right cultural conditions are maintained. During prolonged cultivation *in vitro* however changes sometimes occur in the cells. Some degenerate, and this has been thought to be the result of infection with unknown viruses. Other cells become malignant and thereby attain pathological immortality, for neoplastic cells appear to be capable of unlimited growth whether they are transplanted in animals or cultivated *in vitro*. There is a well known strain of rat tumour which is still being maintained by serial transplantation. The connective tissues cells from which it originated underwent malignant transformation in the reign of Queen Victoria!

The most malignant cancer cells tend to conform to a common cytological type irrespective of their tissue of origin. They have lost their capacity to perform specific functions, and their activities are limited to a continuous self-duplication. Aberrations of mitosis are frequent, and result in irregular distribution of chromosomes to daughter cells, which probably accounts for many cancer cells failing to survive. Since abnormalities of mitosis occur in normal tissue cells, and are numerous in regenerating tissues, it is possible that there is an increasing variation in the growth potentialities of its cells as an organism grows older.

During an investigation into the structure of the cytoplasm under various conditions of growth, I found that mitosis could proceed in the normal manner although the cytoplasmic structure exhibited the characteristic features of degeneration. Cells completed division when their cytoplasm was depleted of nucleotides, but filled with granules and vacuoles, and contained only a few vesicular mitochondria. Mitotic figures were often observed in old tissue cultures in which the majority of the cells were degenerate.

Cytoplasmic structure undergoes a series of changes during mitosis, which is most obvious in differentiated cells in which the mitochondria and Golgi apparatus are confined to certain regions of the interphase cells. At the onset of mitosis the mitochondria tend to become granular and are scattered throughout the cytoplasm, while the Golgi apparatus disintegrates and is also dispersed. During the reconstruction phase the mitochondria increase in numbers and return to their original form and position in each of the daughter cells, and the Golgi apparatus is reconstituted in its particular area. Besides these visible changes, there are concurrent alterations in the sub-microscopic structure of the cytoplasm as is shown by the fluctuations in its viscosity. Thus the cytoplasm, as well as the nucleus, is involved in a cycle of changes which includes a complete reorganization of cellular structure at the reconstruction phase. This might constitute the means of rectifying deterioration of the intracellular mechanism, and restoring it to full functional activity.

SOME ENDOCRINE ASPECTS OF AGEING

By

G. I. M. Swyer

Consultant Endocrinologist, Department of Obstetrics, University College Hospital, London, W.C.1

Ageing is not unnaturally linked with the idea of rejuvenation, and rejuvenation with the sex glands, since loss of gonadal function may be one of the more outstanding general characteristics of ageing. It is of course many years now since the idea was first mooted of bringing about rejuvenation by means of gonadal extracts, transplants or interference, as in the Steinach operations; and it is certainly not without significance that we hear little of these manoeuvres nowadays. Nevertheless such distinct changes in gonadal function, in the human at least, to which species alone this paper will refer, occur with increasing age that ageing, which, in general is a somewhat ill-defined process, can at least be described rather precisely for the gonads. Moreover, recent studies on the excretion of steroid hormone metabolites in older individuals have given interesting information on ageing in relation to adrenal function, so that it may be of some value to review the impact of ageing on the activity of these endocrine glands.

The gonads

Both the ovary and the testis have two distinct functions; to secrete steroid hormones and to produce gametes. The gonadal steroids are not necessarily specific to the gonads themselves, and whereas the ovary, in addition to secreting oestrogens and progesterone (both of which may also be produced by the adrenal cortex) also produces androgens, the testis, in addition to secreting androgens, also produces oestrogen (which may be the " second testicular hormone " or " X " hormone, the postulated existence of which has proved necessary in order to explain certain facts of testicular physiology and pathology). In the testis, androgen secretion is independent of gametogenesis; of this there can be no doubt, since clinically one is very familiar with the picture of the male who is normal from all points of view save that his seminal fluid is devoid of spermatozoa in consequence of failure of spermatogenesis. Some of these men have testes in which the extreme of dissociation between gametogenesis and steroidogenesis is shown; these testes have entirely normal Leydig cells but the tubules contain Sertoli cells only, germinal epithelium being congenitally absent (first described by

Engel, 1947 and by del Castillo, Trabucco and de la Balze, 1947). In the ovary, on the other hand, oestrogen and progesterone secretion are intimately associated with the ripening and shedding of ova, or with the follicular transformations which follow this and produce the corpus luteum.

Furthermore, oögenesis and spermatogenesis show fundamental differences in so far as the former appears to be a wholly foetal process while the latter occurs only in sexually mature males. The actual production of oöcytes takes place in the foetal ovary, under what stimulus is not known, and at birth the ovaries together contain something between 200,000 and 400,000 oöcytes. A substantial proportion of these are destroyed by atresia between birth and puberty. Thereafter, once menstrual function has become established, during each cycle one (or occasionally two) oöcytes mature and are expelled from the ovary. This is accompanied by the destruction of some 20–30 other oöcytes, the primordial follicles associated with them undergoing abortive development during the follicular or proliferative phase of the cycle, and atresia during the progestational phase when the ripe follicle has become a corpus luteum. In this way, there is a continual loss of follicles until eventually the stage is reached when few or none remain. Perhaps 400 will have produced eggs, and nearly a thousand times as many will have served the purpose, not of producing gametes but of secreting oestrogens, since there is little doubt that it is from the follicles developing during the follicular phase (as well as from the corpus luteum, and perhaps from follicles undergoing atresia) that ovarian oestrogens are derived. When all the follicles have been used up, it is clear that the ovary is no longer able to secrete oestrogen—or at any rate, not in the way it was formerly able to do so. It is this physiologically-determined ovarian insufficiency which is responsible for the cessation, either gradual or abrupt, of menstrual cycles which constitutes the climacteric.

After the menopause, the endometrium is still fully responsive to oestrogens and will readily proliferate, even to a pathological extent, and bleed if subjected to sufficiently prolonged oestrogenic stimulation—a matter of common clinical occurrence as the result of ill-conceived oestrogen therapy for climacteric symptoms. The anterior lobe of the pituitary gland is also fully effective and, because the decreased production of ovarian oestrogens has reduced their inhibitory effect upon it, increased secretion of pituitary trophic hormones, especially gonadotrophins, is the consequence. The thyrotrophic and adrenocorticotrophic hormones find responsive target organs and, so it is supposed, these glands may become more or less hyperactive. It is to the combination of lowered oestrogen levels and altered thyroid and adrenocortical function that the symptoms commonly experienced by climacteric women are usually ascribed.

It must, however, be emphasized that postmenopausal women do not generally show evidence of extreme oestrogen deprivation (such as may be found, for example, in Simmonds' disease, the panhypopituitarism of which leads to failure of all the endocrine glands). The source of their oestrogens is therefore presumably the adrenal cortex, and probably also the ovaries, since removal of the latter may be followed by evidence of lowered oestrogen production. Indeed, hyperplasia of the ovarian cortex has been described in postmenopausal ovaries and, admitting the pluripotentiality of ovarian stromal cells, it seems probable that these cells may secrete the postmenopausal ovarian oestrogens (Engel, 1955).

In the case of the human ovary, therefore, we have an easily intelligible explanation of the loss of function with increasing age, leading eventually to cessation of gametogenesis and to the outward consequences of this—the menopause. It may not be amiss to point out that in spite of this cessation of gonadal activity, sexual activity, as registered by the existence of libido and the ability to enjoy normal coitus, persists unimpaired in many postmenopausal women. Clearly the endocrine aspect of sexual ageing in women may be quite independent of the psychic aspect.

In the male, as we have hinted, conditions are somewhat different. Here, not only is steroidogenesis independent of gametogenesis (though normally both occur as a result of pituitary gonadotrophic stimulation) but the production and shedding of gametes which, from puberty onwards, is a continuous process, does not entail loss of primordial germ cells. Apart, therefore, from extraneous factors, both processes could be expected to continue indefinitely until death; and such is indeed generally, though not always, the case. There are many authentic examples of men in the eighties producing fertile semen and begetting offspring. Nevertheless, the testis does show evidence of ageing, though it appears to partake of the more general degenerative process which affects the rest of the body rather than of the quite specific process occurring in the ovary.

The decline in excretion of 17-ketosteroids with increasing age, to which more detailed reference will be made later, points to decreasing activity of the Leydig cells. Actual observations of these in testis sections do not show any obvious decrease in numbers in older men, though various cytochemical changes may perhaps be indicative of impairment of function. The only constant finding in such testes is a progressive increase of tubular fibrosis with advancing age.

Doubts have been entertained regarding the existence of a male climacteric, but it may safely be accepted that there is indeed such an entity, even though only a small proportion of men may be affected. The case has been very well argued by Spence (1954), and it is accepted by all endocrinologists of experience that a small proportion of males do undergo a phase of waning testicular function and may in consequence experience symptoms of the same general character, and no doubt arising in the same way, as are experienced by a far greater proportion of women. There is also general agreement that the male climacteric occurs at a later time in life than the female, usually between the ages of 55 and 65. The first to produce scientific evidence indicating the existence of a male climacteric were Heller and Myers (1944) who found that whereas in 15 men with psychoneuroses or psychogenic impotence the urinary gonadotrophin excretion was normal, in 23 men whom they considered to be suffering from the male climacteric the titre was unequivocably higher. Moreover, testicular biopsies on 8 of these men revealed a reduction in the size and activity of the seminiferous tubules, together with a reduction in the size and number of the Leydig cells, in 5 cases; and in the remaining 3 hyaline degeneration of the tubules. Later observers have reported normal testicular histology, in spite of the presence of increased urinary gonadotrophin; or normal tubules but decreased numbers of Leydig cells with abnormal histological characteristics.

According to Howard *et al.* (1950), the climacteric can be divided into a compensated and a decompensated stage. In the former there is a tendency to decreased production of gonadal hormones, which is met by over-production of pituitary gonadotrophin so that gonadal function remains essentially intact. In the decompensated stage, this tendency to gonadal failure is not counterbalanced, in spite of increased gonadotrophin, and so gonadal failure becomes clinically demonstrable. In the female, the compensated stage lasts only a very short time, but in the male, on the other hand, it is the decompensated stage which is seldom encountered. They therefore considered their patients with the male climacteric whose testicular biopsies appeared normal, in spite of increased gonadotrophin titres, still to be in the compensated stage.

The adrenal glands: changes in steroid metabolism in ageing men and women

The most detailed study of the excretion of steroid metabolites in ageing men and women has been made by Pincus *et al.* (1955) who measured the output of 17-ketosteroids, non-ketonic neutral steroids, reducing neutral lipids (for corticosteroids), androgens and oestrogens in a group of 297 men and 320 women, all considered to be normal and healthy. It is possible only to summarize their findings here.

Oestrogens. In men, the output of oestrogens remains relatively constant with increasing age; in women, on the other hand, the output declines between the ages of 40 and 60 years, reaching a level somewhat below that of men, and thereafter remaining constant. Of the separate oestrogen fractions, oestrone and oestradiol decline slowly in men, accompanied by an increase in oestriol which makes the total oestrogen output appear constant; in women, the most marked decline in earlier decades is in oestriol output, the least marked in that of oestrone, while in the later decades small further declines in oestrone and oestradiol are accompanied by an apparent increase in oestriol. Oestriol is a metabolite, not a secretory product as both the other two may be; its increase with advancing age may therefore be due to lesser destruction of secreted oestrogen.

Neutral steroids. The rate of decline of 17-ketosteroids, as determined both by the Zimmerman and by the Pincus reaction, is similar for both sexes and when the logarithm of the ketosteroid excretion is plotted against age, a linear relationship becomes evident. The urinary ketonic androgens are higher in men than in women and decline more steeply in the former, particularly during the early decades. During these decades, the decline of androgens is steeper than that of 17-ketosteroids, so that with advancing age the ratio of 17-ketosteroid to androgen increases, albeit somewhat irregularly. Since the androgenic activity of the 17-ketosteroids is to be attributed chiefly to androsterone, it follows that the rate of production of androsterone (and presumably of its precursors) declines more rapidly than that of the less androgenically active 17-ketosteroids. Though this might have been expected in men, as a result of declining testicular function, it is perhaps more surprising in women and suggests a decrease in output of either adrenal or ovarian androgens, or both.

The ratio of androgens to oestrogens is higher for men than for women at all decades until the ninth.

The output of neutral reducing lipids, which reflects the production of corticosteroids, is on the average higher for men than for women at all ages, and varies but little with age. In contrast, the non-ketonic steroids, a mixture of substances of doubtful origin, part adrenocortical and part perhaps gonadal, decline with age much as do the 17-ketosteroids. Thus the outputs of the various classes of neutral steroids vary with age in dissimilar fashion. Close study of the data indicates that the steroids of adrenal origin are less affected by age than are those derived from the gonads, but that adrenal steroids are not uniform in behaviour in this respect.

This differential behaviour is clearly shown by the various α-ketosteroids. Thus, the 11-deoxy steroids, androsterone and etiocholanolone, decrease regularly and markedly with advancing age, in both men and women. In contrast, the 11-oxygenated 17-ketosteroids decrease much less markedly with increasing age in both sexes. The 11-oxyetiocholanolones decrease least of all; these substances derive chiefly from hydrocortisone and its metabolites.

Steroidogenesis and ageing. The outstanding fact, referred to above, is that with advancing age the urinary output of corticosteroids changes very little while that of 11-deoxy-17-ketosteroids decreases markedly. This decrease is most probably due to a decline in the secretion of precursor hormones. Now the principal source of neutral steroid hormones appears to be cholesterol, and the two main pathways of its conversion are: (1) splitting of the side chain at carbon 20, leading to the production of Δ^5-pregnenolone, which is the prime precursor of the C-21 steroids; this process is stimulated by ACTH (adrenocorticotrophic hormone), whereas the subsequent reactions by which Δ^5-pregnenolone is converted into hormonal end-products occur independently of ACTH, as do those of the minor pathway to C-21 steroids which does not involve cholesterol; and (2) splitting of the side-chain at carbon 17, leading to the production of dehydroepiandrosterone, which is the prime precursor of the C-19 steroids; it is not known whether there is any pituitary trophic influence on dehydroepiandrosterone production. The fact that corticosteroid production is well maintained with advancing years may be taken to imply that the reactions involving breakdown of cholesterol to Δ^5-pregnenolone, as well as all of the enzymatic reactions leading to the conversion of the latter to hydrocortisone, remain unimpaired. The decline in androgen synthesis implies that the ageing adrenal cortex becomes less able to effect rupture of the cholesterol side chain at carbon 17, since this is the initial step in the production of the C-19 steroids of which the androgens form a part.

Studies of the steroid excretion patterns of ageing individuals given steroid supplements indicate that the capacity to convert these hormonal steroids to characteristic metabolites remains essentially unchanged. This is confirmatory of the hypothesis that the alterations in steroid output with age are primarily due to changes in steroid production and not in the subsequent metabolism of these hormones—with certain reservations mentioned below.

In their most recent report, Pincus and his colleagues (Freeman *et al.*, 1956) made investigations on a group of 14 men, aged 70 to 91, who received daily for five months a steroid mixture composed of testosterone, adrenosterone, hydrocortisone acetate and corticosterone in amounts which, on the basis of previous calculations, were expected to bring up the average level of blood steroids to those found in young men aged 20 to 40. This treatment had no effect upon the serum and urinary concentrations of inorganic ions, upon the blood eosinophil count or upon the excretion of creatinine. The total 17-ketosteroid excretion rose from an average of 4·5 to 17·1 mg. per 24 hours, was maintained at this level throughout treatment, and declined to pre-treatment levels on cessation of therapy. Although the excretion of 11-oxygenated steroids agreed well with the theoretical values, that of 11-deoxy-17-ketosteroids differed from the expected output. Androsterone excretion was 78 per cent higher and etiocholanolone 26 per cent lower than the calculated levels; this suggests that in older men there may be enzymatic changes as a result of which the ratio of androsterone to etiocholanolone derived from the reduction of testosterone is increased, or, as a further possibility, that there is increased conversion of etiocholanolone to etiocholanolone-3α, 17β-diol.

An exercise test demonstrated a significant improvement in muscular strength in 10 of the men, which was maintained to a lesser degree for at least four and a half months after discontinuation of the treatment. On the other hand, there was no change in the psychological status, neuromuscular co-ordination or encephalograms of the subjects—nor were there any untoward effects.

These studies of Pincus and his colleagues, extensive though they have been " ignore a number of hiatuses that may be more significant than the established facts ". For example, they take no account of possible changes in the output of corticosterone, aldosterone, and other adrenal steroids as yet unidentified chemically; nor of other testicular and ovarian secretions. Moreover, knowledge of steroid physiology in older men and women is still very limited. Though the simple notion of certain pioneer endocrinologists that adequate gonad replacement should result in rejuvenation has not received more than minimal support from experimental work, it still remains possible that more complete steroid replacement might not be without value in preventing or counteracting at least some of the general effects of ageing.

References

del Castillo, E. B., Trabucco, A. and de la Balze, F. A. 1947. *J. clin. Endocrin.*, **7**, 493.

Engel, E. T. 1947. *J. Urol.*, **57**, 789.

Engel, E. T. 1955. *Recent Progr. Hormone Res.*, **11**, 291.

Freeman, H., Parsons, O. A., Feffer, M. H., Phillips, L., Daneman, E. A., Elmadjian, F., Bloch, E., Dorfman, R. I., and Pincus, G. 1956. *J. clin. Endocrin.*, **16**, 779.

Heller, C. G. and Myers, G. B. 1944. *J. Amer. med. Ass.*, **126**, 472.

Howard, R. P., Sniffen, R. C., Simmons, F. A. and Albright, F. 1950. *J. clin. Endocrin.*, **10**, 121.

Pincus, G., Dorfman, R. I., Romanoff, L. P., Rublin, B. L. and Bloch, E. 1955 *Recent Progr. Hormone Res.*, **11**, 307.

Spence, A. W. 1954. *Brit. med. J.*, **1**, 1353.

DISCUSSION

J. B. Hamilton. The studies by Pincus and his associates extend and confirm much previous work by other investigators employing assay and chromatographic methods. These studies are in essential agreement with one another and with results from other fields of interest in showing that a progressive decline in 17-ketosteroids, androgens and oestrogens occurs after young adulthood.

Contributions to urinary steroids by the adrenal cortices are not always readily distinguishable from those by the gonads, and so my colleagues and I have investigated urinary titres of 17-hydroxycorticosteroids in something over fifty surgically-produced healthy eunuchs, 19 to 70 years of age, and in a comparable series of intact men. There was no evidence of compensatory increase in adrenocortical function after gonadectomy, so that these subjects appear to provide acceptable information uncomplicated by testicular secretions. Titres of 17-hydroxycorticosteroids, which in man are chiefly metabolites of hydrocortisone, did not change significantly with age in either castrate or intact men. In contrast, values for 17-ketosteroids per kg. of body weight did decrease with age in eunuchs as well as in intact men as has been reported previously. Extra-testicular contributions to 17-ketosteroids seem to be considered by most students to arise largely from the adrenal cortices; to the extent that this belief is correct, the decline in 17-ketosteroid titres of ageing eunuchs represent a decrease in secretion of their precursors by the adrenal cortices.

DEMOGRAPHIC ASPECTS OF AGEING

By

B. BENJAMIN

General Register Office, London, W.C.2

THE title of this paper can be regarded as applying to two distinct problems which are far too often confused, viz., first the ageing of the population structure, second the ageing of individual people. The most common misconception which has greatly bedevilled consideration of social services and especially of both dependency and employment in old age, is that the contemporary increase in the relative numbers of elderly people is itself evidence of improved vitality and increased longevity of those currently reaching normal retirement age.

A few simple facts may put matters in perspective:

(1) As yet (an important reservation) the increase in the proportion of elderly people in the population has been much less affected by mortality changes than by first a rise and then a fall in the annual flow of births during the past century.

(2) In the first half of this century[1] the proportion of male babies surviving to their fifth birthday increased from 79·4 per cent to 96·2 per cent; the proportion of five-year-old boys surviving to age 65 increased from 49·5 per cent to 68·4 per cent; but the proportion of 65-year-old men surviving to their 70th birthday increased only slightly from 76·1 per cent to 79·8 per cent. The expectation of life of a man aged 65 increased from 10·80 years to 11·73—a relatively modest improvement (though, of course, *any* extension of life at extreme ages is a real advance).

The ageing of populations

It will be advantageous to consider first the changes in the age structure of the population that have taken place in recent decades, since these changes have given rise to social and economic problems which have impressed themselves on the public conscience to the extent of exciting interest in the welfare of the elderly person as an individual. It is doubtful whether there would be the same interest that there is to-day in the process of ageing as such were it not for the problems, impressively large in dimensions, of dependency and chronic invalidism associated with these population changes.

[1] This is a broad statement covering a comparison of mortality in England and Wales in 1950–52 with that in 1901–10.

The population pyramid

It is useful to illustrate the age structure of a population by means of a pyramid (see fig. 1) in which numbers in successive age groups are represented by the areas of horizontal strips placed on top of each other with the youngest at the base and the oldest at the top (each strip is split into left and righthand portions proportionate to the numbers in each sex.) They form a pyramid because as a result

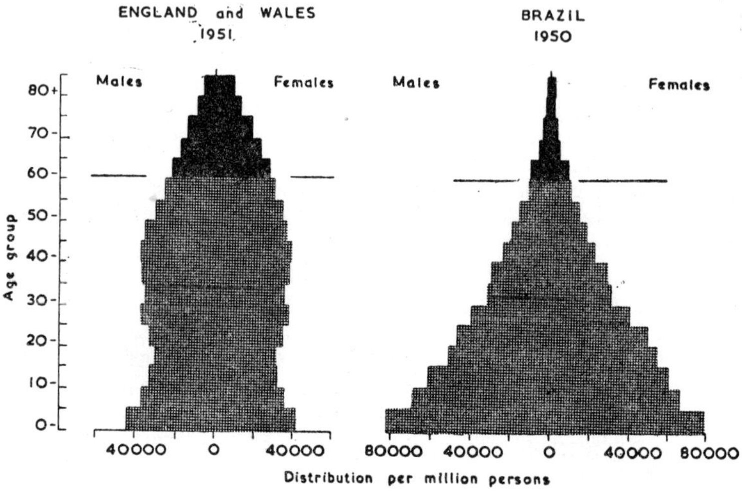

FIG. 1.—Population Pyramids

of deaths the numbers remaining in successive age groups become smaller and smaller until a point is reached at which there are no survivors and this is the peak of the pyramid.

The proportion of older people in the population, say those aged 60 and over, is represented by the part of the total area of the pyramid which is at the top above the strip which begins at age 60 (dark hatching of fig. 1). In a country with very heavy mortality few people will survive to older ages and the pyramid will be short and flattened—the right-hand pyramid relates to Brazil, an area with high fertility and high mortality. If mortality is low then a high proportion of the people will survive to older ages and the pyramid will be comparatively tall with more steeply sloping sizes—the left-hand pyramid relates to England and Wales in 1951.

Successive age group strips in the pyramid represent successive generations, the youngest or most recent being at the bottom.

Changes in fertility

Any change in the intake into the population by births will have an immediate effect on the lowest strip in the population pyramid but will only affect the older strips at later periods of time as the affected generations survive to the relevant older age groups. Fig. 2 shows what would be the shape of a typical pyramid 5, 15, 35, 55, 95 years after a sudden 20 per cent fall in annual births. It will be seen that

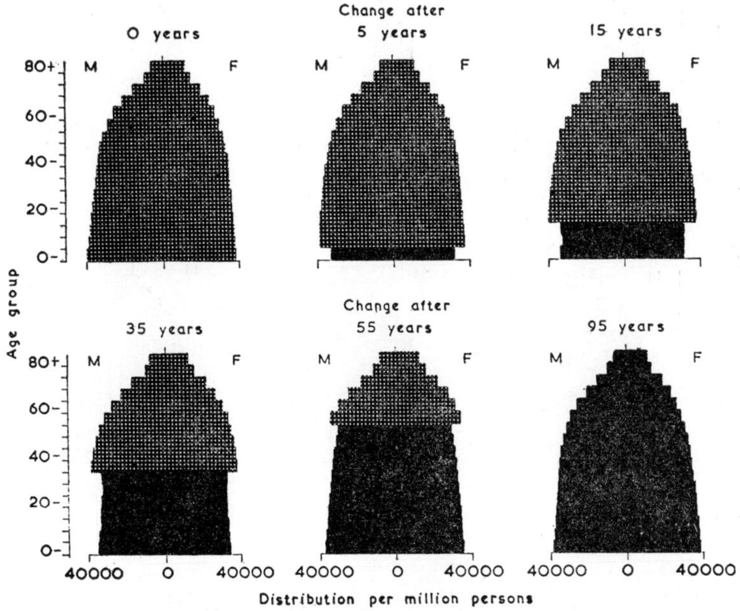

FIG. 2.—Change in pyramid following fall in births

the base of the pyramid is almost immediately contracted and then the contraction slowly moves up the pyramid. Until the contraction reaches the older age group that group represents an *increasing* proportion of the total population. Eventually when the pyramid is wholly based on the changed birth intake its original shape is restored. This is important, viz., that the effect on age structure of a decline in births is immediate; ageing is prolonged but the change is nevertheless not permanent.

Changes in mortality

A change in mortality will increase the size of those strips in which deaths are postponed. If mortality is reduced at all ages the effect will be permanently to stretch the pyramid upwards and thus to increase the proportion in the older age groups. If however the reduction is only at younger ages this will increase the slope of the sides of the lower part of the pyramid but will not affect the top until sufficient time has elapsed for the increased survivors at younger ages to reach the older age groups. Fig. 3 shows the shape of a

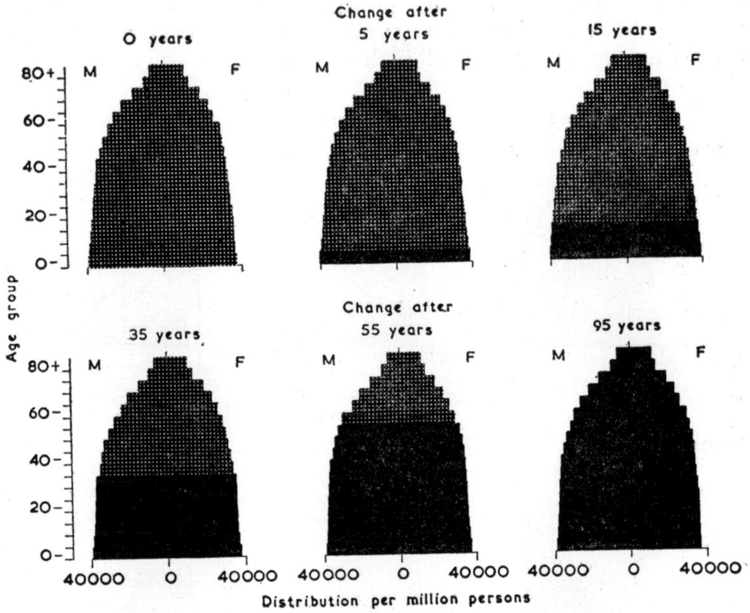

FIG. 3.—Change in pyramid following fall in mortality

typical pyramid 5, 15, 35, 55 and 95 years after a 50 per cent reduction in mortality under age 30 (no reduction above that age). It will be seen that ageing is not so immediately apparent but though deferred is a permanent end result.

The actual changes in England and Wales

The changes that have been taking place in the age structure of the population of England and Wales have been in fact a mixture of the kinds of changes depicted in figs. 2 and 3.

Live births in England and Wales in successive decades from 1841 were:

1841–1850	5,488,736	1891–1900	9,155,153
1851–1860	6,471,650	1901–1910	9,298,209
1861–1870	7,500,096	1911–1920	8,096,222
1871–1880	8,588,782	1921–1930	7,129,070
1881–1890	8,890,238	1931–1940	6,064,516

It will be seen that up to the decade 1901–1910 births were increasing in numbers. The pyramid was broadening at its base (fig. 4). The reverse process of fig. 2 was taking place. The population structure was becoming youthful. In 1911 a stable population

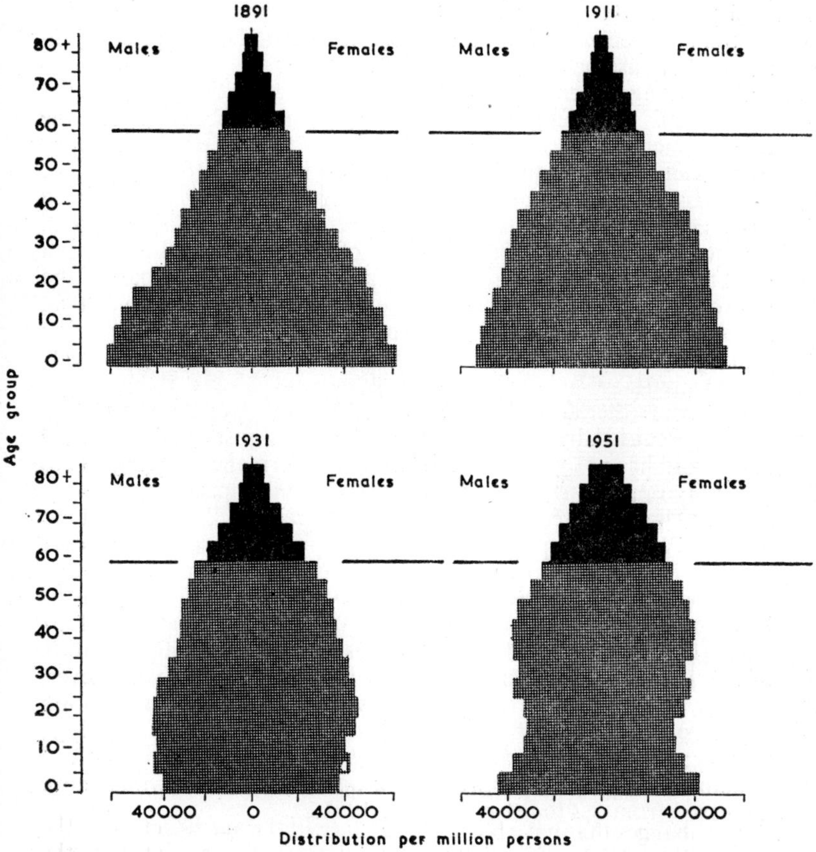

FIG. 4.—England and Wales. Actual population pyramids. 1891–1951

sustained by constant births and subject to constant mortality—a " normal " population—would have had 10·4 per cent of persons aged 65 and over. The proportion was actually 5·2 per cent. Thus when the fall in births began to take place, the population was abnormally young and even if births had remained constant it would have had to " grow up ".

This process of " growing up " was accentuated by the decline in births. The base of the pyramid became narrower and narrower, and a bulge was created which has now moved up to the higher age groups (fig. 4). If births remain relatively constant the bulge will cease to have any appreciable effect on the proportion of older people in the population by the end of the present century.

Side by side with these changes mortality has been steadily declining especially at younger ages at which the conquest of infectious disease has been most effective.

At most ages there have been considerable reductions in mortality since 1910. At younger ages there has been an acceleration in the improvement in survival rates in the last twenty or thirty years. The effect of this reduction in mortality has been to increase the steepness of the sides of the population pyramid up to quite advanced ages, i.e. to make it run up to a point less rapidly than it would otherwise have done. The full effect of this change will not be felt until those generations which have experienced the major part of the improvement in mortality, viz., those born in the last decade or two, have reached old age. This will not be until the end of the present century (*after* the effect of the past decline in fertility has largely disappeared).

It appears therefore that the temporary ageing of the population structure caused by fertility changes has only just reached its full extent and has not yet spent itself while the permanent ageing due to mortality improvement has yet to make its full impact.

It may be estimated that if mortality had remained constant at the level of 1910–12 the proportion of the population aged 65 and over at 1951, in England and Wales, would have been 9·7 per cent, compared with an actual proportion of 11·0 per cent. If annual births had remained constant at the level of 1901–10 the population in 1951 would have numbered 51·5 millions (instead of an actual population of 43·8) of which 9·4 per cent would have been aged 65 and over. On the other hand we have already seen that the 1911 population structure was already abnormally youthful as a result of earlier high fertility; this means that even if *both* mortality and fertility had remained unchanged the proportion aged 65 or more would have increased from 5·2 to 8·1 per cent in 1951.

We may set out these changes as follows:

Per cent 65 and over in 1911	5·2
Increase due to abnormal structure in 1911 (i.e. due to the prior rise in births)	+2·9
Increase due to fall in births since 1911	+1·6
Increase due to fall in deaths since 1911	+1·3
Per cent 65 and over in 1951	11·0

Up to the present therefore the total fertility changes have made a much greater contribution to the increase in the proportion of older people in the population, than has the decline in mortality.

The ultimate effect of the changes

These underlying trends have been made the basis of a forecast of the future population structure of England and Wales by the Government Actuary in consultation with the General Register Office. According to the most recent projection based on the population at mid-1955 (General Register Office, 1956) it has been estimated that on certain stated assumptions as to the future course of fertility and mortality, the population of England and Wales in 1995 will number 46,328,000 of which 7,477,000 or 16 per cent will be aged 65 and over.

There will be some changes at younger ages, though these will be comparatively small. Over the next twenty years the population between ages 15 and 65 as a proportion of the total population will decline only slightly from 66·1 per cent to 65·1 per cent; and in 1995 the proportion will have been reduced to 64·7 per cent. The proportion of young persons of working age, viz., 15–34, which was 27·4 per cent in 1951 is expected to be 28·0 in 1975 and 26·4 per cent in 1995. Numbers of males aged 15–64, which have increased from 13,052,000 in 1931 to 14,062,000 in 1951, will have reached 15,191,000 in 1995. In 1931 50 per cent of men of these ages were under 35; in 1951 42 per cent; and in 1975 and 1995 the percentages are expected to be 44 and 41 respectively. (The temporary rise in the proportion in 1975 is associated with the large numbers of births which took place in the immediate post-war years). On the whole it can be said that the working population has " grown up " and has attained comparative stability in its age structure.

Dependency and economic activity

Older persons who have retired from employment have always consumed goods and services, whether purchased by their own savings, by relatives or by social welfare funds (national or local). An increase in numbers of older persons does not create a situation which is new but it does bring into sharper relief an economic fact of great importance in relation to social accounting, namely that the social product created by those who work is currently shared with a dependent population who are beyond working age (as it is also shared with children who are too young to work).

It is possible to assess this factor of transfer of claims on consumption goods from workers to dependents and the trend of the economic pressure it constitutes, by considering the ratio of dependents to workers in the past, present and the future.

Subject to some reservations we may consider as elements of the dependent section of the population (1) children under age 15 (2) non-gainfully occupied housewives, etc. (3) persons who have attained the minimum retirement ages (for National Insurance purposes) of 65 (men) or 60 (women). The balance may be regarded as the supporting section of the population. It is proposed to estimate the non-gainfully occupied housewives, etc., in 1975 and 1995 by taking 60 per cent of the female population 15–59.

We then obtain the figures shown in the table below.

TABLE I.—The economic pressure of the dependent population in England and Wales, 1901–1955 (numbers in thousands).

Year	Total population all ages	Children under 14 (up to 1931) 15 (1951–)	Non-gainfully occupied females 14–59 up to (1931) 15–59 (1951–)	Pensionable class i.e. males 65+ females 60+	Remainder	Ratio to (6) of Col. (3)	Ratio to (6) of Col. (4)	Ratio to (6) of Col. (5)	Ratio to (6) of Total of cols 3, 4 and 5
(1)	(2)	(3)	(4)	(5)	(6)	(7)	(8)	(9)	(10)
1901	32528	9878	6565	1998	14087	·70	·47	·14	1·31
1921	37887	9772	8188	2972	16955	·58	·48	·18	1·24
1931	39952	8909	8521	3842	18680	·48	·46	·21	1·15
1951	43745	9733	8376	5990	19646	·50	·43	·30	1·23
1975[1]	46364	9219	8149	8462	20534	·45	·40	·41	1·26
1995[1]	46328	8885	8106	8742	20595	·43	·39	·42	1·24

[1] Based on the population projection referred to on p. 59.

Several interesting features emerge from this table. The first is that the total pressure of national dependency (as indicated by Col. 10) has not changed substantially compared with 1921, a rise in pensioners being offset by a fall in child dependency (considering education and other subsidized services it is arguable that, from a national point of view rather than that of the family, a child should be reckoned to require the same relative amount of food and services as an adult) while the dependency of adult women has been a stable element. For the future it appears that pensioner dependency will in the next forty years rise by some 40 per cent but almost all of this rise will be offset by a fall in the dependency of children and unoccupied women in so far as the two kinds of dependency are capable of being equated, a point on which it is difficult to give a clear answer. In 1995 children and unoccupied women will be fewer and workers slightly more numerous, on the assumption of the population projection. The *total* national dependency will not be much greater in 1995 than in 1951.

One demographic aspect of the growing pressure of dependency is the restraint which it places upon the emigration of younger active persons upon whose productivity the elderly population will in future depend.

The ageing of individuals

There are difficulties in the way of extending the working life. Although mortality has been considerably reduced—to the extent that the expectations of life at age 65 on the England and Wales mortality of 1954 were: males 12·0, females 15·0; compared with 11·3 and 13·1 in 1930–32—it is not easy to interpret this increased vitality in terms of increased capacity to work. Postponement of death in an elderly person by improved medical care does not imply the arrest of the relentless process of degeneration though doubtless raised standards of health have generally retarded this process.

We are brought face to face with the second of the two problems we distinguished at the very outset of this discussion, viz. the ageing of individuals as distinct from the ageing of the population structure. An immediate handicap is the lack of any comprehensive statistics of the characteristics of the elderly population especially in relation to their general fitness for employment and their special aptitudes for particular types of employment.

Fig. 5 shows death rates for males in England and Wales during the period 1950–52 for individual ages above 30. It will be seen that mortality is increasing quite steeply even before age 60 and that after that age the rise is much accelerated. These figures provide no clue as to the optimum retirement age.

The Fifth Interim Report of the Government Actuary on the National Insurance Act 1946 (1956) indicates that weeks of sickness benefit per annum per employed male in 1949–52 rose from 1·64 at 40–44 to 2·97 at 50–54, 4·17 at 55–59 and 6·11 at 60–64. Further statistics supplied by the Ministry of Pensions and National Insurance show that the proportion of the insured male population who were sick on 5th June, 1954, and had been sick for more than three months rose from 1·73 per cent at ages 45–49 to 2·73 per cent at ages 50–54, and at individual higher ages the percentages were:

55–	4·33	60–	6.64
56–	4·61	61–	7·88
57–	5·30	62–	8·81
58–	5·48	63–	10·22
59–	6·51		

The proportion begins to rise rapidly after age 60 and even before age 65 more than one-tenth of insured men are in poor enough shape to have been on sickness benefit for more than three months.

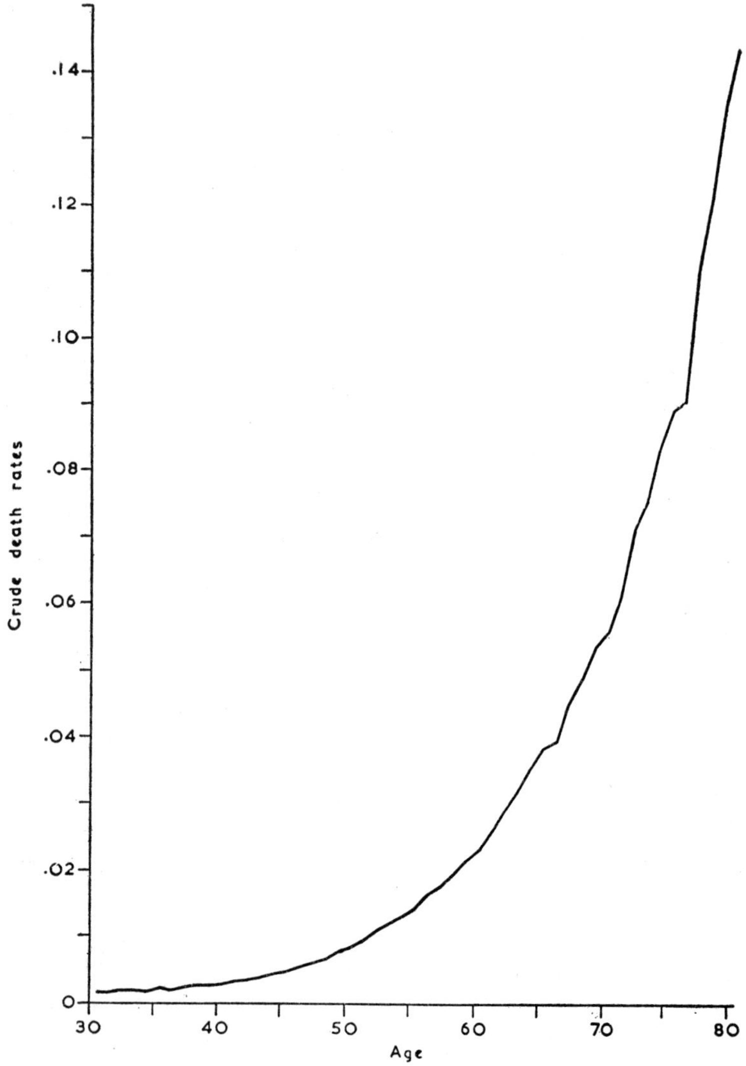

FIG. 5.—Crude death rates for England and Wales, 1950–52 males

A limited study of general practitioner records carried out by the General Register Office and covering ten practices during 1953–54 indicated that the percentage of male patients who consulted their doctor at least once in the year rose from 59 at 15–44 and 45–64, to 66 at 65+. In a single practice (Fry 1952) it has been found that while the average attendances per year for all patients was 3·3, for old people in the age groups 60–69, 70–79, 80 and over the numbers were 4·1, 7·0 and 7·7.

Clearly the problem of invalidity at ages even younger than the present normal retirement ages is still considerable, and though the upper limit of working life may indeed be much above age 65 for men, the population still working at older ages becomes more and more highly selected as the age advances. The social and medical services which are more concerned with the sickness and retirement experience of the general population than with the tenacity of the hardier members will face a growing burden of chronic invalidity arising from the sheer growth in numbers of older persons outstripping the slower improvement in their average vitality.

On the other hand surveys of the aged population have indicated that infirmity is more quickly developed and more passively accepted in conditions of stagnation and boredom and is more effectively resisted and prevented by interest and occupation.

The extent of employment

There is indeed no reason why there should be passive acceptance of invalidity either by the individual or by the community. It may well be that the proportion of workers who pursue employment beyond normal retiring age is not as large as it could be. At the 1951 Census when the effect of National Insurance was to impose a virtually compulsory retirement age of 65 upon a large proportion of workers, the proportions of men gainfully occupied at older ages were as follows:

Age	*Per cent of males at each age, gainfully occupied*
55–59	95·0
60–64	87·5
65–69	47·2
70–74	27·4
75 and over	12·6

At the 1921 Census when there was some economic depression but no widespread system of contributory pensions to influence

F

retirement the proportions employed beyond age 65 were considerably larger, viz.:—

55–59	94·0
60–64	88·7
65–69	79·4
70–74	52·6
75 and over	27·0

Whether this is evidence of the sustaining influence of occupation, or whether it merely represents the less definite transition from full-time working through stages of unfitness to retirement which existed before contributory pensions were introduced is difficult to judge. The real problem which the statistics do not at present measure is the extent to which the economy of the country fails to match the limited and specialized aptitudes of older workers to equally limited and specialized occupations so that they may make the maximum possible contribution to the national product within their physical capacity and enjoy doing it. There is a great need for really scientific and objective information about aptitudes—to this the family doctors may hold the key.

Housing and family life

There is one last demographic aspect of an ageing population which cannot be left out of account, viz., the family units in which older people live, especially in relation to the incidence of loneliness, and their housing conditions. Tabulations of the 1951 Census (One per cent Sample) show that in Great Britain in that year there were 2,198 thousand households with married heads age 60 and over[1] and about half of these households, 1,129 thousand, consisted of two persons mostly but not quite all of them married couples living alone. Of 3,558 thousand unmarried[2] persons enumerated over the age of 60 about one-quarter, viz., some 904 thousand were living alone. There were also some 625 thousand unmarried heads aged 60 and over of households of only two persons. Finally account must be taken of a considerable number of unmarried persons aged 60 and over enumerated in non-private households, i.e. in hospitals, nursing homes, mental institutions, homes for the aged or infirm, and in larger hotels and boarding houses (those containing at least 10 rooms); these number nearly 200,000.

The present picture is therefore of about a million married couples with the head aged 60 or over living alone, over a million and a half unmarried persons aged 60 or over, either living alone or with one other person and some 200 thousand in institutions and hotels. These are large numbers and they seem likely to grow larger.

[1] Ninety-seven per cent of all married males aged 60 and over were enumerated in the status of head (or spouse of head of) of household.

[2] Single, widowed and divorced.

References

Fry, J. 1952. *Brit. med. J.*, **2**, 249.

General Register Office. 1956. Quarterly Return of the Registrar General for England and Wales, 4th Quarter, 1955, No. 428.

General Register Office. 1956. Studies on Med. Pop. Subj. No. 9. General Practitioners' Records.

Government Actuary. 1956. National Insurance Act, 1946, 5th Interim Report, HC 274.

DISCUSSION

R. A. M. Case. I should like to raise a point of terminology. The curve of death rates in fig. 5 bears the legend " crude death rates " but the curve shown appears to be one of age-specific sex-specific death rates. Most demographers reserve the term " crude death rate " to a rate that does not take age into account and it appears to me to be confusing to introduce the term here with a different meaning.

B. Benjamin. The National Life Table for 1950–52 is under construction by the Government Actuary and the rates published therewith will be corrected to remove artificial sources of irregularity. The rates used in my diagram were the straight ratios of deaths 1950–52 to census populations (multiplied by three) without any adjustment. In order to distinguish these from the official life table rates and to deter any tendency to take the irregularities of the curve at their face value I used the term " crude ". I thought that since the curve showed age rates there would be no confusion; but I am sorry there was no space in my severely pruned paper to explain this.

N. W. Pirie. If a medical board is considered competent to judge whether a scientist is fit to take up a job in the tropics, and this decision may greatly affect his salary, why should not a suitably constituted board decide whether he is fit to continue work or should be retired? There does not seem to be any need to have a regular retiring age at all, we should all be considered on our merits.

As a further example of a man being retired while still fully competent there is Sir Charles Martin who, on retiring from the directorship of the Lister Institute, took on the job of reorganizing the scientific research of Australia. There would be no great difficulty in establishing the board competent to decide whether retirement of an old man is necessary; with a little supplementation it already exists in the staff of the institute that he directs.

B. Benjamin. There is a case for saying that old people should have the opportunity to consider their own " merits "; taking medical advice but not having it imposed upon them.

N. W. Pirie. Is the increase in the proportion of women who get married a consequence of a change in the sex ratio or of the progressive removal of the disabilities from which married women have suffered in jobs such as teaching or the civil service?

B. Benjamin. It is difficult to apportion the increase between these two factors because people do not register their reasons for choosing a particular age at which to marry. It seems likely that the change in the sex ratio has improved the marriage prospects of younger women but it is also clear from the persistence of high marriage rates that these prospects are being taken advantage of more readily and at younger ages by both sexes.

P. Townsend. The two ideas of " dependency " and " support " in old age are fundamental to discussion of the economic " burden " of old age. In Table I Dr. Benjamin shows past and future trends in the economic pressure of the dependent population. As dependents he includes children of school or under school age, non-gainfully occupied females and people of pensionable age. The question of the dependent population is a tricky and important subject. In the first place, future population estimates are continually being revised. During the War for example Beveridge used an estimate of the elderly population in the 1980s which was $1\frac{1}{2}$ million higher than recent official estimates for the same years, and until recently the whole discussion of the economic problem of the aged has been conducted on the assumption that a decreasing number of workers are to carry a burden of an increasing number of old people. Dr. Benjamin has helped to dispel this mistake, but the matter is more complicated than he has described. He counts the minority of old people still at work, I think mistakenly, in the class of dependents, and he does not say whether any assumptions should be made about the unemployed and the sick. Are they dependents? If so, the picture of the trend since the 1930s will look very different. And the matter does not rest here if one wants to get the facts straight. The number of domestic servants, for example, has decreased by a million, and presumably many housewives have taken over the same work, yet why call the latter dependents and not the former? There are many other difficulties in sorting out what one means by the " dependent " population. The housewife *supports* her husband so that he can go to work. What I am saying is that most workers are supported by dependents. Old people often support their relatives as much as do young wives their husbands. In an investigation carried out in East London recently two-thirds of grandmothers over the age of 60 were found to have the regular care of at least one grandchild. The grandmother helps her married daughters and sons living in neighbouring streets just as they help her. In the last twenty years in Britain there has been an increase of over a million in the number of married women at work. I believe this increase is related to the increase in the number of grandmothers in the population. The upshot is we need to be careful about our statements of the changing number of dependents in the population and how far old people depend on or support others.

B. Benjamin. I quite agree with Mr. Townsend. The statement in my paper was over-simplified and devoid of all qualifications only on account of the need to keep the paper short. It was only possible to introduce a rough guide to current trends. A fuller and more balanced discussion will be found in " The growth of pension rights and their impact on the national economy " (Bacon, F. W., Benjamin, B., and Elphinstone, M. D. W. 1954. *J. Inst. Actuaries*, **80,** 141), and in the " Report of the

Committee on the Economic and Financial Problems of Provision for Old Age " (1954. Cmd. 9333, H.M.S.O.).

E. Digby. Mr. Benjamin spoke very wisely when he pointed out the enormous advantage which would accrue to the State finances, to national production and to individual health if even so little an extension as one additional year could be added to the active working life, averaged over the whole population.

Reference has been made elsewhere to the greater survival rate of women but I believe that the general impression is correct that though women last longer they seem to age more completely while they are doing so. From both these aspects it seems a pity that 95 per cent of gerontological research workers are men and almost as big a proportion of their work is done on male age-decay. I hope that something will be done to remedy this state of affairs since women, whose biological life histories are so largely concerned in reversionary changes, should in consequence be better adapted to the major reversionary changes than are men.

R. A. M. Case. I think it might be interesting to consider age-specific death rates at different times from a standpoint that is complementary to the actuarial viewpoint taken by Dr. Benjamin. The technique of cohort-analysis can be regarded as a biological approach to the situation. For those present who may be unfamiliar with this method I will try to show the principle on a simple diagram. (Figure 6.) If we set up an animal experiment at a point that we call zero time and date, and make observations on any attribute (we shall consider death rates only) on the survivors of the original batch of animals at regular intervals of time, we can enter the observations on a grid where one axis is marked in intervals of time

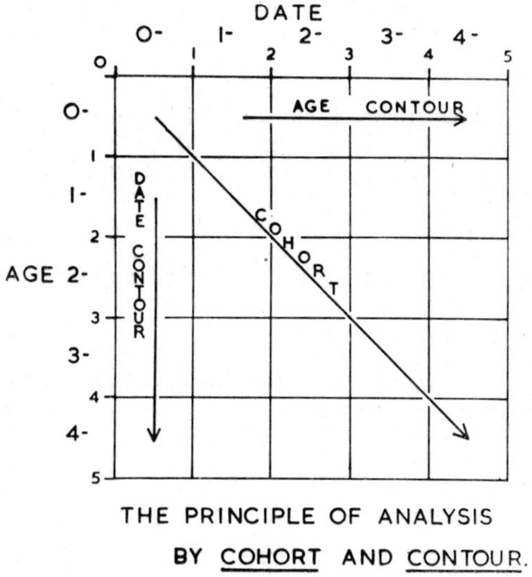

Fig. 6

(i.e. date) and the other in equal intervals of time added to the starting age of the animal. If for simplicity we assume the starting age to be zero also, this axis will be age.

It is now obvious that observations relating to the original batch of animals will fall along a diagonal (marked " cohort ") of this grid. If separate series of experiments are started at regular intervals, all the cells of the grid can be filled with observations, but each individual experiment is described only by the diagonal. If each batch of experimental animals is kept under different environmental conditions, we should not regard the record found in a vertical column as a biological statement of the mortality experience of the type of animals used, since each successive observation is derived from an experiment carried out by definition under different conditions.

The changing human environment may be likened to such a series of experiments. Those born in 1871, who are now represented in the entries under the last age group in the Registrar General's returns, lived through very different conditions during the first twenty years of their life from those who were born twenty years ago, and had very different therapeutic measures available to them at each age. In other words, they were a batch of animals kept under different conditions.

The form of social accountancy that circumstances compel the Registrar General to adopt each year means that the series of age-specific death rates that he computes represent the vertical series (date contour) on the grid, and therefore cannot be considered as a biological statement of the mortality experience of any group of people. The expectation of life at any age is a summarizing index calculated from this vertical column, and probably has little biological meaning whilst environment is still altering.

In Figure 7 I show how the age-specific death rates have changed in England and Wales in the last century when viewed by cohort. (The figures are drawn from a quinary-quinquennial grouping and the cohorts labelled with the central birth-date of the group contributing to the life-experience described). In order not to overload the figure, cohorts spaced twenty years apart have been selected. The progressive improvement in mortality experience is shown, and it is now clear that to extrapolate these curves, which is what must be done to obtain an expectation of life from this method, would be nothing but guessing.

I should like to make it clear that I do not claim that one statistical method is " right " and the other " wrong ". The relevance at any time must depend on the question to be answered, and the Registrar-General is not always concerned with the biological approach in his annual accounting.

Miss E. A. M. Bradford. In considering the supposed financial burden of the aged it is well to remember that, as time goes on, with the contributory pensions system people will, in effect, be paying their own pensions and the aged are really self-supporting. If this were not so they would be paid a proportionately higher wage during their working life from which they would save for their old age as in previous generations. Pensions, as now arranged, are not a charity to the aged given by a benevolent population but a convenient means of compulsory saving for those who cannot, or will not, save during their best financial period of life.

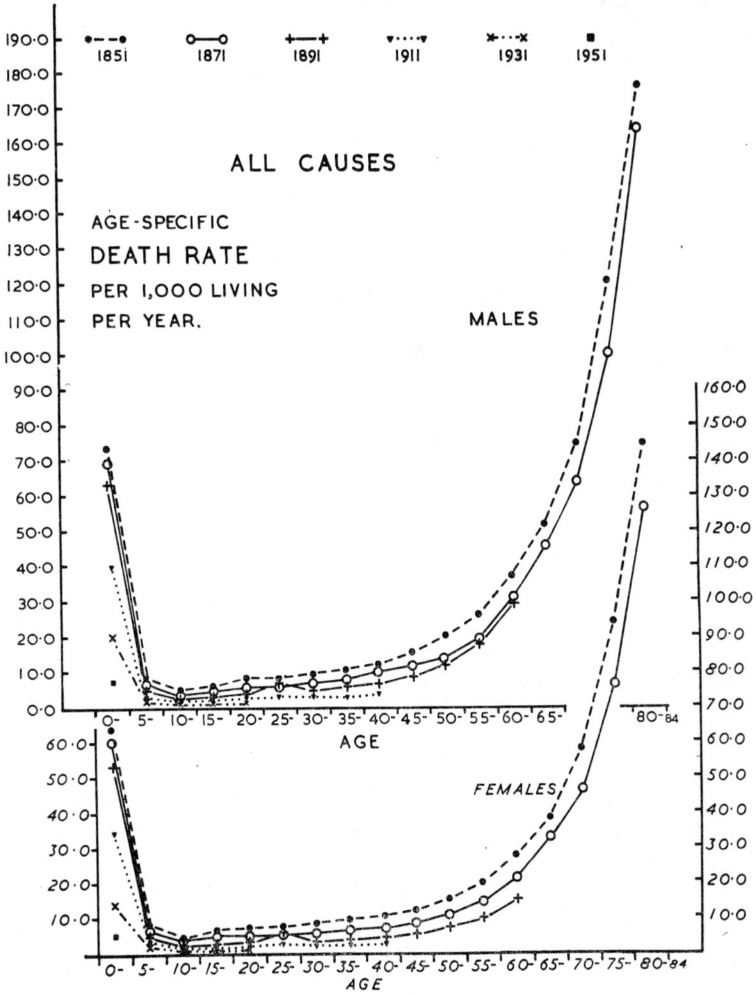
1851
1871
1891
1911
1931
1951
ALL CAUSES
AGE-SPECIFIC
DEATH RATE
PER 1,000 LIVING
PER YEAR.
MALES
FEMALES
AGE
AGE
190·0
180·0
170·0
160·0
150·0
140·0
130·0
120·0
110·0
100·0
90·0
80·0
70·0
60·0
50·0
40·0
30·0
20·0
10·0
0·0
0- 5- 10- 15- 20- 25- 30- 35- 40- 45- 50- 55- 60- 65- 80-84
0- 5- 10- 15- 20- 25- 30- 35- 40- 45- 50- 55- 60- 65- 70- 75- 80-84

Fig. 7

FUNCTIONAL CHANGES WITH AGE IN RELATION TO THE EMPLOYMENT OF THE ELDERLY

By

W. HOBSON

Department of Social and Industrial Medicine, Sheffield University

JUST as children present structural, functional, metabolic, immunological and personality characteristics peculiar to their age, so do the aged. In the case of children there is a fairly sharp dividing line between childhood and adolescence at the age of puberty, but there is no sudden change in ageing; the process is a gradual one.

From the official point of view, old age is taken to be the age at which contributory pensions are first payable, that is, sixty in the case of women and sixty-five in the case of men. But the onset of physiological old age is not synonymous with any chronological age; it varies a great deal from individual to individual, because of the many variables in mode of living, accidents, disease, and hereditary factors.

Whilst it may be difficult to define old age as far as the individual is concerned, it is sometimes easier to define ageing in relation to the properties of individual cells or systems. We know that in some cases ageing of cells begins from a very early age, even before birth, but such cells are not essential to life or proper functioning of the whole organism. In general there tends to be cellular atrophy, fatty infiltration and increased pigmentation. There is an increase in connective tissue and a decrease in tissue elasticity, particularly in the skin and blood vessels. Bones tend to become decalcified and calcium to be deposited in other situations, such as arteries, thyroid, pleura, cartilage, tooth pulp, valves of the heart and scar tissue. These changes may be partly due to a general decrease in the blood supply of tissues resulting from generalized arteriosclerotic changes.

There is also an inefficiency of the homoiostatic mechanisms which maintain the constancy of the physico-chemical equilibrium; thus, there is a great vulnerability to changes in temperature, and metabolic and vasomotor responses are less effective. There is less efficiency in coping with sugar changes (Smith and Shock, 1949), whilst many elderly people fail to respond to an injection of insulin by a fall in blood sugar (Himsworth and Kerr, 1939). Small variations in temperature and pulse have a much greater significance in the elderly. There appears to be a wide range in the resting pulse rate, and in a person with a normal resting pulse of 50 to 60 a pulse rate of 74 may represent a considerable degree of tachycardia. Symptoms and signs are less conspicuous than in younger persons and minor deviations take on a greater significance. Repair is much slower and there is a narrower margin of safety; thus anaemia is much more detrimental to the arteriosclerotic person.

The rise in blood pressure with advancing age is well known, but it is much more difficult to say when this becomes pathological. It is clear that many old people can remain perfectly healthy with a raised blood pressure. Hobson and Pemberton (1955) found, for example, that in a random sample of old people living at home, 43 per cent had a resting diastolic blood pressure of 100 millimetres of mercury or over and the majority of these were in good health.

In the central nervous system there are a number of changes which, although in a younger person they would be of clinical significance, appear to be of little importance in the elderly. Thus, muscular weakness, brisk knee jerks, wasting pupils which do not react to light, and loss of vibration sense, can all occur in the absence of any evidence of organic disease. Similarly, there are certain mental changes which are characteristic of old age, such as impaired memory for recent events, which may make it difficult to take an accurate history. Apathy and depression have important effects on behaviour.

There can be little doubt that an adequate well-balanced diet is an important factor in preserving the health of the elderly. Mental factors characteristic of the elderly such as apathy and forgetfulness may adversely influence health by leading to a monotonous and inadequate diet. In addition, atrophic changes occur in the digestive tract which can have marked effects upon the absorption of nutrients; for instance, achlorhydria is present in 35 per cent of elderly people over the age of sixty years (Bloomfield and Polland, 1933), and the concentration of pepsin and also tryptic activity fall off markedly in later life (Meyer, Spier and Neuwelt, 1940). Recent surveys have established the high incidence of chronic atrophic gastritis over the age of fifty years. It is well known that dyspepsia and constipation are common. These various factors are no doubt responsible for some of the wasting or malnutrition that occurs in old age, or, in severe cases, for deficiency diseases such as scurvy and anaemia.

Changes in the skeletal system can have an enormous effect upon a person's physical capacity although they might be quite unimportant as a danger to life. Examples of these are arthritis and defects of the feet, both very common. Rheumatoid arthritis, for example, limits greatly the skilled work that can be performed by the hands and is a great disability to a musician or typist or to the housewife. Osteoarthritis and spondylitis of the spine are both very crippling deformities for a labourer. The tendency to decalcification is a particular danger when falls occur.

Changes in the special senses are very important in determining capacity for employment. Deafness, especially for the higher tones, is quite common, increasing with age and more common in men. There is some evidence that this may be due to acoustic trauma in industry (Hobson and Pemberton, 1955). Vertigo, on the other hand, is more common in women and probably associated with arteriosclerosis of the labyrinth. This can be most distressing, leading to falls and fractures, and is an important cause of impaired mobility (Droller and Pemberton, 1953). Visual acuity and the quickness of visual perception decrease, whilst the minimum amount of light required for visual stimulation increases. These changes can be detected at the age of twenty years. Some experiments by Weston (1949) indicate that the loss of visual acuity with age may be more severe than is revealed by the commonly used tests.

If the accommodation of the lens measured in diopters is plotted against age, the values all fall within two smooth curves which enclose the normal range for accommodation and thereby set up the standards of normality at the different age groups. The power of accommodation falls steadily with age until the age of fifty years when it remains stationary. We can say, with a fair degree of accuracy, that in the great majority of people physiological senescence in the human lens is reached at the age of fifty years, but it begins before puberty.

Until recent years, we have had little information on the normal ranges of many of the biochemical variables, and even the values for younger age-groups are often based on inadequate figures. Thus serum cholesterol, serum alkaline phosphatase and blood urea levels are all higher than in younger age groups.

Physical Capacity

As far as employment is concerned, however, it is the functional capacity of the whole individual which is important. The amount of physical work that can be performed by the elderly person has been studied extensively in the laboratory. These studies, however, are full of snags and do not necessarily apply to conditions as found in occupational situations. It is well known that after maturity, the speed reaction time and strength of skeletal neuromuscular mechanisms are decreased. In moderately hard physical work there is very little decrease in the mechanical efficiency in older persons, apart from those over seventy years of age. In severe work, however, there is a marked diminution in this ability with age; the chief physiological factor limiting performance appears to be the inability to supply the increasing amounts of oxygen required by the tissues (Robinson, 1939).

There is a definite deterioration in the speed of performance in the higher age-groups; the changes in women take place sooner and are more marked. There is a similar decrease in dexterity. Older men who have had mechanical training appear to be able to maintain the speed rates of young adults (Miles, 1954). This lends experimental proof to the well-known observation that experience and practice can counteract the increasing disabilities of old age. Smith (1938) made a study of men of different age-groups working under conditions similar to those found in factories and he confirmed the observations of Robinson that in short periods, or with moderate work, the performance of the elderly was comparable to those of the younger age-groups. When, however, the conditions became arduous, there was a definite decrease in the ability to perform high-speed manual work. In all age-groups, however, there are great variations in individual ability, so that some of the older workers were able to better the performance of some of the younger workers. This is an important consideration and emphasizes the principle that individual differences should be taken into account whenever the capacity of elderly persons is being considered.

Especially important in old persons is accident proneness. De Silva (1938) found that many of the components involved in the driving of a car showed a steady deterioration over the age of thirty years but the mileage driven per fatality showed an improvement with increasing age, at any rate up to the age of fifty years. Vernon, Bedford and Warner (1928, 1931), in their studies on the accident rate in coal miners, found a decrease in frequency up to the age of fifty years and that increasing care and experience can counteract the decreasing performance which might be predicted on the basis of physiological factors alone. The figures for absenteeism are of interest in this respect but the results as one would expect are conflicting since they are biased by the fact that only those able to maintain good attendance records remain employed.

Intellectual Changes with Age

A considerable amount of work has been carried out on the relation between intelligence and age and changes are found similar to those noted in sensory and motor efficiency, particularly where speed is concerned. This decrease in the score achieved in the intelligence test takes place near the age of fifty years. Again there are wide individual differences in every age-group. The smoothed curves indicating the decrease with age in intelligence test scores take the form of a parabola, with a rapid rise up to the age of fifteen years, reaching a peak about the age of twenty years, and a gradual fall after the age of fifty years, with a more rapid decline after the age of seventy years.

The ability to learn depends largely upon the intelligence, so that memory function, ability to learn new tasks and tasks involving the relinquishing of old habits are found to be more difficult for old people. Welford (1951) has carried out some experiments of a rather different kind. These involved tests of logical thinking in a group of subjects engaged on university teaching and research. The experiments, although not conclusive, showed that the older subjects tended not to draw logical conclusions based strictly on the statements given but often confined themselves to general remarks upon the statements. In the older learner interest in the subject aids the mental organization necessary for attention and retention. Whereas in the younger person, active, varied learning is the rule, the older person's learning is concentrated more on some particular field in which the person is interested. On such tests as vocabulary and general information older adults do as well as younger cases, at least among those who are better educated. This may be due to education itself or to higher intellectual capacity. There are some suggestions that in the course of education at a university level, people acquire certain intellectual skills which do not desert them (Owens, 1953). A special intelligence test administered at the time of college entrance and administered again thirty years later showed that scores are on the whole higher on the re-test.

It would appear that although many age changes are continuous in nature, their effects are not necessarily so. A certain minimum capacity is required for performance of any kind and so long as the age changes are not sufficient to bring mental or physical capacity below that needed, then performance will be adequate. Once capacity drops below this minimum, there may be a sudden fall in performance. The two critical age-periods appear to be fifty and seventy.

There is a great need for further research in this field and we require more information on the actual occupational capacity of different age-groups. The selective process which weeds out the more unfit with increasing age, complicates studies in this field and indicates a need for longitudinal follow-up studies of groups of individuals in younger age-groups.

Chronic Sickness in Old Age

So far most of the changes in mental and physical capacity which have been described are what one might term physiological, but in the elderly there is a great deal of chronic sickness which the subject must suffer for the rest of his life. Hobson and Pemberton (1955) found that about 29 per cent of males and 45 per cent of females suffered from some pathological condition which impaired mobility. Most of this was due to cardiovascular disease (48 per cent of those disabled), joint conditions (13 per cent of those disabled) and chronic bronchitis (13 per cent of those disabled). Other important conditions were general weakness, vertigo, mental state and impaired vision. It would appear that a high percentage, at least one-third, of all over 65 suffer from impaired physical capacity as a result of chronic disease.

Age and Retirement

The evidence would seem to lend no support to the idea of fixing sixty-five years of age for retirement from employment rather than, say, seventy. In view of the wide variations in abilities found in the older age-groups it might be thought desirable to have some means of discrimination by psychological tests at the age of sixty-five years, although this might well be considered undesirable on other grounds. There are certain physical conditions which would disqualify a person from continuing work after the age of sixty-five years, for example, severe angina, a previous stroke or failure of vision. It is important to remember that those with the best educational and cultural background show the least change with age and any diminution in the powers of critical reasoning may be more than compensated by the knowledge and wisdom gained as a result of experience. One thing is certain, and that is that there is a great need for more refined tests which will measure intellectual capacity in the aged. The present tests do not measure many of the higher mental abilities and tend to discriminate against the elderly, particularly where speed is a factor. They take no account of difficulties due to physical disabilities, for example, deafness, tinnitus, defects of vision, tremor, weakness.

Although prior to the war the percentage of elderly workers in whole-time employment was falling, it would appear that this trend has now been halted; at any rate in 1954 there were employed in this country, 870,000 men and women over pensionable age and 230,000 over the age of seventy years. Fleming (1955), in a survey of the age composition of the iron and steel industry has shown that there are considerable differences in the age structure of firms carrying out similar work, some firms making fuller use of older manpower than others.

The problem has also concerned the non-industrial worker. The retiring age for professors at Edinburgh University is seventy years, whilst Cambridge and Oxford Universities recently raised the retiring age from 65 to 67 years. In an institution for higher education in New York City, ten professors with a total age of 706 years have been brought out of retirement in order to undertake teaching. It is to be noted that in many occupations, for example: Members of Parliament, clergymen, craftsmen, and in private practice in the professions, the date of retirement is in many cases well over seventy years of age.

According to Fleming, what is required is, ". . . an approach which includes one or several of such industrial procedures as a periodic medical check-up of such persons, job analysis and work study applied in this connection, later-age counselling or retraining, or definite and deliberate fitting of jobs to men." One thing is certain; it is quite impossible to make any but the very broadest generalization in regard to employment. Occupations are so very diverse and individuals, as we have seen, vary so much in their capabilities, destinies and desires. We certainly require much more research in the basic sciences before we can even begin to apply these Utopian principles with any degree of confidence, and the doctor must still rely largely on his so-called clinical awareness and intuition in advising on the employment of the elderly.

References

Bloomfield, A. L. and Polland, W. S. 1933. *Gastric Anacidity: Its Relation to Disease*. New York.
De Silva, H. R. 1938. *Sci. Mon., N.Y.*, **47**, 536.
Droller, H. and Pemberton, J. 1953. *J. Laryng.*, **67**, 689.
Fleming, C. 1955. *Old Age in the Modern World.* London; E. & S. Livingstone.
Himsworth, H. P. and Kerr, R. B. 1939. *Clin. Sci.*, **4**, 153.
Hobson, W. and Pemberton, J. 1955. *The Health of the Elderly at Home.* London; Butterworths.
Meyer, J., Spier, E. and Neuwelt, F. 1940. *Arch. intern. Med.*, **65**, 171.
Miles, W. R. 1954. *Geriatric Medicine*, edited by E. J. Stieglitz, 3rd edition, London.
Owens, W. A. 1953. *Genetic Psychology Monographs*, Massachusetts.
Robinson, S. 1939. *Arbeitsphysiologie*, **10**, 251.
Smith, K. R. 1938. *J. appl. Psychol.*, **22**, 295.
Smith, L. E. and Shock, N. W. 1949. *J. Geront.*, **4**, 27.
Vernon, H. M., Bedford, T. and Warner, C. G. 1928, 1931. *Rep. indus. Hlth. Res. Bd. Lond.*
Welford, A. T. 1951. *Skill and Age.* London; Geoffrey Cumberlege, Oxford University Press.
Weston, H. C. 1949. *Trans. Illum. Engng. Soc., N.Y.*, **14**, 281–297.

DISCUSSION

T. Levitt. I was very intrigued to note the high values attained for cholesterol in normal men (409 mg. per cent) and normal women (481 mg. per cent). The values for alkaline phosphatase were similarly high (up to 21 units for normal men and 27 units for normal aged women).

How was normality assessed in the older people? Was it possible to exclude latent or subclinical disease? Were all these patients studied for a number of years subsequently to exclude this possibility?

I would suggest that it would be still more helpful if these " normal " values were separately assessed for the various decades, say 50 to 60 years, 60 to 70 years and so on.

W. Hobson. The survey was carried out on a random sample of old people over pensionable age living at home, either alone or with spouse. Everyone showed some evidence of disease; I consider that any person who did not at this age should be classed as being decidedly abnormal. We excluded any cases which were suffering from a disease which was a recognized cause of alterations in the serum cholesterol or alkaline phosphatase levels, for example, in the serum cholesterol group we excluded one case of disordered thyroid function and three cases of glycosuria, and in the alkaline phosphatase group we excluded eight cases of Paget's disease. We are at present carrying out a follow-up survey so I cannot answer the question or whether any of those with high levels subsequently developed a disease recognized as a cause of high value in these biochemical estimations. It is of interest to note, however, that in one of the cases which had a high serum alkaline phosphatase, an X-ray carried out four years later for another condition which had developed showed evidence of Paget's disease and there was no evidence of this, of course, in the original survey. The mean level of serum cholesterol decreased after the age of 70. The mean level in a control group of younger people was 211 mgms./100 ml. for men and 233 mgms./100 ml. for women, considerably lower than the levels for the older age group. As we did not do any estimations on people in the age group 50–60, I cannot give any values for that decade.

THE WORKING LIFE AS A MEASURE OF AGEING IN MEN AND ANIMALS

By

F. Le Gros Clark

The Nuffield Foundation—Research on Ageing within Industry London, N.W. 1

It is well to remember that for working men or draught animals the first sign that irreversible organic changes are taking place may be the simple fact that they can no longer carry out their normal tasks in the accustomed way. The tasks they are set to carry out provide a measure of organic changes that may affect their endurance, the pace of operations, output, decision, "nerve" and any other characteristic that has to be tested against hard experience. We can call the process deteriorative if we like; but it is only deteriorative inasmuch as the ageing men or animals concerned have hitherto been expected to reach a certain daily standard of performance under fairly constant conditions. Adjustments in their method of working, or in the design of machines and vehicles, could conceivably have enabled them to continue longer on the job; as matters stand they are ceasing to be fully employable.

In this paper I merely offer a few suggestions of ways in which the working lives of men and animals could be used as a measure of the characteristics and comparative rates of what we conventionally call ageing. Traditionally the daily work has been so planned that healthy men and animals may be expected to cope with their appropriate tasks until they reach some recognized chronological ages. Because of the wide range of individual variations these accepted chronological ages have necessarily to be expressed as lying somewhere within a term of years. Though the age selected (such as that of 65 years for working men) is an arbitrary one, it has some relation to practical experience; for in practical experience few men, and for the matter of that few working elephants, would be reckoned as indubitably ageing till they had reached their late fifties. What then concerns them (and their employers or owners) are not the organic changes that would no doubt have been physiologically traceable through their earlier working years; for these changes would so far have had little or no recognizable bearing upon their working efficiency. What concerns them is the plain evidence they now have of an irreversible and possibly a progressive decline in their ability to make the grade in comfort.

In studying the effects of ageing among men and animals it is not advisable to make any assumptions based on the observed tolerance shown by younger subjects for heavy work, heat, humidity, long hours, etc. The evidence from these sources may or may not be relevant if we happen to be dealing with ageing subjects who have been for many years exposed to the same conditions. It would be more profitable to examine patiently what factors are applicable to ageing subjects, possibly indeed to ageing subjects alone. In other words, we need a type of job analysis that takes account of the physical and psychological changes we suspect to be characteristic of advancing years. But we have one great advantage. The tasks carried out by an ageing man or animal are usually " constant " in the sense that they have probably been on the same jobs for most of their working lives. Any decline in their effectiveness or staying power is thus measurable or it can at least be described in objective terms. The results of a study of this kind have always to be reduced to a statistical form, because the tasks undertaken are so varied and the characteristics of ageing differ so widely from one subject to another. If we carry our investigations far enough, certain common patterns and common factors should begin to emerge. Though all research of this kind is necessarily laborious, our consolation is that we may discover ways of easing and possibly prolonging the useful working lives of innumerable elderly men; and this would often be of far greater moment to the men themselves than the mere knowledge of an approaching dissolution.

I am more especially interested in the human problem, not only because I am human but because it seems to me that we have reached a stage in our affairs at which all records of the working lives of draught animals are going to become progressively less available to us. Industrial and agricultural undertakings that may formerly have kept reliable records are replacing horses with tractors and motor vans; the old records have often been long since destroyed; and annual redundancies among horses are making current records, such as they are, virtually useless for estimating the average working life. The literature of the horse is curiously lacking in statistical information on the subject of ageing. For the most part we have to rely on the memories of experienced men. The range of ages up to which a horse may continue on the work for which he had originally been bred or acquired is naturally a fairly wide one; but an age of 10–15 years is commonly stated to be the average. Ensminger (1951) says that " the market value of the animal increases rather sharply with maturity and then decreases beyond eight years of age. On the other hand, for many purposes horses are quite useful up to twelve years of age or even longer." It appears that in London the railway yards would normally have been acquiring horses at about six years of age; and the medium-built trotting horse used in the light parcel vans would then have been expected to remain on the job for an average of six further years or maybe less. It might then be transferred to slower and possibly heavier work if its health and general conformation allowed; or it might be disposed of for relatively light work on a farm. I am informed that a Corporation horse or a goods horse on the railways employed on heavier work but at a slower pace might, after having been acquired at six years, enjoy an average working life of a further eleven years or more.

It is impossible, however, to discuss the subject in brief, because much would depend on the hours and conditions of work. When men become legally entitled to work shorter hours, so usually do their horses; and the working lives of horses may then be prolonged by the resulting avoidance of colic, undue fatigue and strained tendons. The market value of a horse is some indication of the years of useful work expected from it; but of course over a term of six to ten years it may provide no evidence on its day-to-day performance as a draught animal. Among horses that survive the hazards of injury, disease and destruction, the expectation of life from birth is commonly recorded as being one of twenty-five to thirty years. In that case the comparative brevity of its reported working life seems to merit some research. Healthy animals that have been disposed of for light work in towns or on a farm seem usually capable of a few hours' work a day for a varying number of years.

With regard to the working lives of timber-hauling elephants, Departmental Instructions for Forest Officers in Burma (1936) state that for purposes of valuation mature animals would be reckoned as those from 21 to 45 years of age. Those exceeding 45 years are defined as " aged ", and thereafter 5 per cent of the original value should be written off annually. This suggests that the extreme useful life of an otherwise healthy elephant would not far exceed 60 or 65 years. Williams (1950) says, " We may assume that the elephant has on the average a working life of from its 20th to its 55th year "; and Evans (1910) mentioned that Burmese forest men with considerable experience of elephants account an animal of 35–45 years as fully grown, one of 45–50 years as " middle-aged " and one of 50 to 60 years as " after mid-day " or in decline.

Since elephants have been known to live beyond 70 years, those that survive their period of useful service may be turned out " to grass ", foraging for themselves in the forests. But wastage due to disease, accident or overwork would evidently on many occasions have been quite high. Benedict (1936) quotes a statement that of 78 elephants over 20 years of age in the possession of the Madras Forest Department only 7 were over 50 years and but one over 60. Milroy (1922) remarks that wrong hours of work and insufficient feeding are the two chief causes of elephants falling sick; and in reference to old age he says that it may cause the forearms to lose their great development of muscle and the legs to become thin.

Williams (1953) records the occasional use of " old crock elephants " (presumably those over 55 years or more) in pushing and lifting logs with the head, tusks and trunk from the sand banks of rivers into the main channel. This is described by him as " a form of work reserved for pensioned tuskers suffering from old age, deformities or incurable wounds that made them unfit for harness ". As for the final cause of death in old age, Evans (1910) says that death may occur suddenly " from long-continued and unobserved disease of the heart. The heart on post-mortem is sometimes enormously enlarged and may rupture from extreme exertion ". He adds that in old animals degeneration of the blood vessels may predispose to rupture, and that this may be brought about by violent or prolonged exertion. A further description given by Williams (1950) suggests that he considers deterioration of the heart a fairly normal cause of death in aged elephants. " His cheeks are sunken, his teeth worn out. Gathering his daily ration of 600 lbs of green fodder has become too great a tax on his energies, and he knows he is losing weight. Old age and debility slowly overtake him."

In a personal letter Lieut.-Colonel J. H. Williams has written, " I would say that an elephant of 50 needed particular watching. He was placed in the company of younger animals, that could take the biggest strain, the longest haul and the heaviest timber. . . . If he lost condition at all at that age, it was doubly difficult to put it on again. . . . The first sign of age really affecting the animal's work was loss of power in his legs. In other words, he looked as if he must sit down to work rather than stand. He gave the impression that, if his head and his trunk could be used without taxing his strength by climbing the mountain side, then he held his own. . . . For this reason, at 60 years of age, generally speaking, the animal continued his work on the flat with his tusks, trunk and forehead. . . . I have known many elephants that continued until they were 70, and they were mostly on that work from 60 to 70 years of age. Between 65 and 70 they often showed signs of giddiness, slightly staggering if kept long hours ".

The ageing man in industry has naturally been more closely studied in recent years than have horses or elephants. But no attempt will here be made to review the literature of the subject. I shall confine myself to a few examples of research procedures that seem to be yielding good practical results.

To demonstrate briefly how the statistics of occupation, health and age may be combined for the purpose of research, I will ask what proportions of working men are likely to have become more or less incapacitated for further work by their mid-sixties. In most manual occupations men do not retire by choice before they reach the pensionable age of 65. Men have nevertheless to pass into retirement before that age if they are compelled to do so by injury or sickness. According to the 1951 Census (1 per cent sample), the percentages of men aged 60–64 in various large occupational groups who recorded on their census forms that they were by that age already in retirement, ran as follows:

TABLE.I.—Percentages of Men aged 60–64 who had already retired from the Occupations in which they are Classified.

Occupation	*Percentage already in retirement*
Building Workers: (Bricklayers, Plasterers, Labourers, etc.)	5·0
Painters	6·2
Farm Workers	6·7
Textile Workers...	6·8
Plumbers...	7·1
Workers in Wood: (Cabinet Makers, Carpenters, etc.)	7·8
Dockers	7·9
Coal Miners	10·1

Though Census figures of this type have to be handled cautiously, it seems probable that most of the men who here said they were in retirement had in fact left work altogether and had not merely transferred to some alternative job. The high figure among coal miners is presumably accounted for by the special hazards of the industry. For the rest, in such broad occupational groups as these we may evidently expect a normal wastage of 5 to 7 per cent even before the men have reached their mid-sixties; and that this is due in great measure to ill health or the strain of work is suggested from other statistical sources. In a four weeks survey of the men who reached the age of 65 in September–October, 1953, the Ministry of Pensions and National Insurance found that about 10 per cent of the men of that age had been incapacitated through sickness for six months or over. In June, 1954, the percentage of men aged 60–64 reckoned as certified incapable of work for over six months was 6·9—a figure closely resembling the retirement percentage of men of that age derived from the Census Tables of Age and Occupation. The proportion of long-term invalidity of this kind seems to increase year by year, from the mid-fifties upwards, reaching a level of 8·9 per cent of all the men at risk at the age of 63[1].

While statistical evidence suggests that some of the men shown as retired before their mid-sixties had probably left the labour market several years earlier through injury or chronic sickness, it seems yet certain that the ailments of old age increase steadily from the mid-fifties onwards. By the time they are due for a State pension, at least 10 per cent of the men may probably be reckoned as more or less beyond further work. It would not be so easy, however, to estimate what proportions of the working men of 65 are beginning by reason of age to show less marked signs of unmistakable wear and tear. Such men have usually to modify the pace or load of their work; or they have to be granted some concessions in the hours worked or the part they play in a labouring gang, or they have to seek transfers to some alternative and less exacting jobs. This is indubitably the critical problem in all studies of industrial senescence; and researches on it are being diligently pursued in various countries. It is possible perhaps by way of illustration to quote a few provisional figures.

In view of the obvious importance for the older men that arrangements should be made for modifying the load of their work, it is essential to know under what working conditions they are most likely to need adjustments of this kind. From the available evidence I am inclined to agree for the moment with Belbin's tentative conclusions (1955). " Transfers ", he says, " tended to occur from the early 50's from operations which required continuous bodily movement and activity in association with a rapid tempo of work, accentuated by the payment of piece-rates or enforced by the requirements of a working team, a series of machines, or a conveyor-line. A few workers on these operations, however, were able to continue to the retiring age. On other operations older workers were rarely found to move to other work, except where there were indications of failing health. Heaviness alone, i.e. with no further strain imposed by the tempo of work, did not appear to cause moves with increasing age."

In other words, it is the relative pace or unrelieved intensity of the work that seems first to tell upon an ageing man. Provided he is given opportunities for adjusting the mode and timing of his work progressively to his age, there is probably little reason for supposing that an experienced healthy old labourer could not accommodate himself in comfort to most of the strains imposed by any heavy lifting, excavating, etc., he has to undertake. But Belbin does not mention that the various hazards, emergencies and responsibilities that a man may occasionally encounter in the course of his work seem also to play a significant part in deciding at what age he has to quit his customary occupation.

[1] It will be observed that the figures here quoted refer to sickness of longer duration than the parallel sickness figures that have been used by Benjamin in his paper " Demographic Aspects of Ageing ".

A study of industrial records usually enables us to arrive at a provisional estimate of the rate of wastage in old age, provided the records have been kept with sufficient care. As an example of provisional estimates of this kind, I reproduce a table that was based on small samples of old building workers, bus drivers and bus conductors. The contrasts are suggestive of the significance of the job itself in determining at what comparative ages men have to leave their various occupations. The records were taken without bias from among the records of men who had recently left employment and were traceable to the last moment of service with the undertakings in which they had been employed. Some allowances were made as far as possible for the relatively few cases of men who had not been removed solely by reason of old age, sickness or death; but it must be admitted that in some instances the entitlement to a State pension had probably tipped the scale for an ageing or ailing man in favour of retiring. I may mention that all the building workers in my sample were employed on maintenance and repairs; and they probably had a better average chance of surviving late on the job than they would have had under the more arduous conditions of constructional work.

TABLE II.—Rate of Industrial Survival as affected by Age, Ill health and Death.

Age	**60**	**63**	**65**	**66**	**67**	**68**	**69**	**70**
Building Workers (320) per cent still on job	100	89	83	75	63	55	46	32
Bus Drivers (150) per cent still on job	100	86	78	50	31	21	15	7
Bus Conductors (150) per cent still on job	100	85	80	35	26	23	16	11

It must be noted, then, that the various groups of workers were all traced from the time they reached or were approaching the age of 60. In each case the figure of 100 represents the men who had been at that age apparently fully effective employees; the percentages for successive years represent the men who had not yet departed from their occupations by reason of age, sickness or death. They were in fact the industrial survivors from a gradually diminishing cohort. Whether those who were thus discharged or retired from their customary jobs found some light alternative work, in most cases we do not know; but a number of them certainly made an effort to obtain further employment.

The decline in numbers over the mid-sixties is not invariably, and perhaps not even largely, due to a tradition of retiring at 65 years. In several industries men seem to delay the moment when they go in search of a less arduous job until they have the security of a pension on which to fall back; and they incline thus to procrastinate in spite of increasing strain or disability. Cumulative evidence from a variety of sources suggests that in many occupations at least 20 per cent of the men are compelled by their mid-sixties to moderate or change their form of work if they are to have any prospect of remaining in settled employment. This 20 per cent lies over and above the 10 per cent of men who have probably by that age been forced through chronic ill health into final retirement. The proportion for whom concessions become unavoidable increases thereafter roughly with the ages of the men, though in some occupations (e.g. bus driving or work at a coal face) few allowances could be made for a man's age.

What I am suggesting, then, is that the working lives of animals and men offer us a chance of glimpsing, and even in some degree measuring by a reliable yardstick, the physical changes that indubitably characterize old age. Under favourable conditions of care and health working animals could provide us with the better subjects for study, because animals presumably have less regard for conventional ideas about old age and about the proper age of retirement. But, as I have said, we have possibly come too late in history to use draught horses and hauling elephants for research of this kind. The study of men under industrial conditions will involve us in Work Studies of a very special type, in which we shall have to distinguish what varied factors in the relation between the ageing men and their jobs are making it necessary for the men either to moderate or to change their form of work. We should have to avoid ready-made theories and presuppositions. The method of research recommends itself, because in our ignorance of what happens in senescence we should at least be observing men just as old age was undeniably overtaking them.

References

Belbin, R. M. 1955. " Older People and Heavy Work." *Brit. J. industr Med.*, **12**, 311.

Benedict, F. G. 1936. *The Physiology of the Elephant.* The Carnegie Institution, Washington. 62.

General Register Office. 1952. *Census 1951 Great Britain. 1 Per Cent Sample.* London; H.M.S.O. Tables II. 3 and II. 6.

Clark, F. Le Gros. 1954. *The Later Working Life in the Building Industry.* London; Nuffield Foundation.

Clark, F. Le Gros. 1956. *Bus Workers in their Later Lives.* London; Nuffield Foundation.

Departmental Instructions for Forest Officers in Burma. 1936. 153.
Ensminger, M. E. 1951. *Horse Husbandry*. The Interstate Printers and Publishers; Danville, Illinois. (Sect. on " The Importance of Age ".)
Evans, G. H. 1910. *Elephants and their Diseases*: a treatise on elephants. Superintendent of Government Printing; Rangoon. 2, 229 and 256.
Milroy, A. J. W. 1922. *A Short Treatise on the Management of Elephants*, Shillong Government Press; Assam. 4 and 32.
Ministry of Pensions and National Insurance. 1954. *Reasons Given for Retiring or Continuing at Work*. London; H.M.S.O. Table 4.
Williams, J. H. 1950. *Elephant Bill*. Rupert Hart-Davis Ltd.; London; 51 and 83.
Williams, J. H. 1953. *Bandoola*. Rupert Hart-Davis Ltd.; London; 55.

DISCUSSION

W. E. Kershaw. I and my colleagues have for some time been interested both in the longevity of parasites and of their insect vectors, and in the survival of a property which may be termed the " infective potential " of the parasite and of the corresponding property of the vector, i.e., its capacity to support the development of the parasite. Our own results, together with those of Dr. Nicholas in Liverpool and Dr. Hopkins in Glasgow, show that these two sets of functions are related and that they follow much the same course as the survival of other animals in controlled conditions. In some experiments however, it is difficult to determine whether the logarithm of the rate of mortality increases with age, for variation in the results obtained becomes greater with the increase of time.

In attempts to explore the relation between death and the survival of biological functions other than death, we have examined the working life of groups of police-horses and draught-horses, the milking life of Ayrshire and Friesian cows, and the breeding life of the sow. Mr. Le Gros Clark has pointed out that such criteria are arbitrary, and that the results of such studies may be of uncertain accuracy and therefore misleading. We were aware of this possibility, but to our surprise we found that the curves obtained were smooth, and that the survival of the animals, as assessed on these arbitrary economic criteria, was such that the logarithm of the rate of mortality increased in a linear relation with age. Histograms plotted from the frequency of the economic disposal of the animals with time proved to be of two types, some being normal and others log-normal in their distribution. We are at the moment unable to understand why this difference should occur and what its significance may be.

W. Wilson Mayne. There are two remarks I should like to make. The first arises from Mr. Le Gros Clark's fears that the time is fast approaching when it will be almost impossible to secure information on ageing from studies on working animals. While this may be true in Britian it is otherwise in India where agriculture remains largely dependent on draught animals. It is perhaps not inappropriate to indicate the opportunity offered for such studies as those suggested by Mr. Le Gros Clark. Draught cattle are widely used at the numerous agricultural experimental farms in India, and this, together with the strong religious prejudice against killing useless cattle, affords material eminently suitable for such studies. Investigations initiated on the farms could be continued at the goshalas being developed all over India as asylums for cattle which have passed their accepted working life.

The second point arises from the discussion on the propriety of a fixed retiring age. While from the biological point of view there is much to be said against fixing a retiring age, any form of social insurance would appear to make such fixation essential. We are faced in the tea industry in India with the introduction of a social insurance system involving a statutory provident fund and retrenchment and retirement compensation but since it is at present virtually impossible to terminate any worker's employment, apart from senior executives, I find myself bound to support the fixation of a definite retirement age. While naturally biological factors must be taken into account, the reasons for fixation are not so much biological as financial and economic.

PSYCHOLOGICAL CHANGES WITH AGE: THE PRESENT STATUS OF RESEARCH

By

ALASTAIR HERON

Medical Research Council Group for Research on Occupational Aspects of Ageing, University of Liverpool

IT is perhaps important to make it clear from the outset that the " psychology of ageing " is always in need of definition, lest it be misidentified as the " psychology of the elderly ". This tendency to think in terms of the problems presented by the elderly is of course perfectly understandable: there are a great many old people these days, most of whom are in need of care, of understanding, and of workable solutions to their difficulties. Nevertheless, here as elsewhere in the biological sciences pure and applied, it is likely that prevention will in the long run prove better than cure; if this be agreed, then it is the *process* we must study, rather than the end-product.

As any organism develops, it changes; as man is a relatively complex organism, the changes will naturally be difficult to assess. More importantly, as the time-span within which these complex changes occur is long (becoming longer), and as the object of study belongs to the same species as those who study it, the difficulties are made even greater. It is therefore not surprising to find such recent statements as that of Arnhoff (1955): " While there is considerable literature available regarding the psychiatric and psychologic changes occurring with advanced age, little agreement exists as to what these changes are or what they mean." The purpose of the present review is to gather together some of the more satisfactory research material in such a way as to indicate points of departure for fresh enquiries.

Beach (1954) has suggested that three generalizations may be made about all aspects of organic development. Of these the first two are sufficiently familiar to be amusing: " Development always depends upon certain crucial external forces which act upon the developing system . . .", and ". . . development is equally dependent upon inherent characteristics of the system "! The third, however, is relevant to our purpose: " An inherent sensitivity to environmental influences is not constant, but rises to a maximum at a particular age and then decreases."

The outstanding feature of our present picture of the developmental process is certainly one of failing potentialities, and it might be tempting to regard Beach's generalization as but a different way of saying just this. There is however the other aspect, allowed for by Beach, but not so often noticed in the general literature on ageing: a " decrease in sensitivity to environmental influence " may well represent a psychological asset—but only if we are prepared to regard man as more than an input-output system. The outstanding work of Lehman (1953) leaves us in little doubt that, so far as the creative intellectual fields of endeavour are concerned, both quality and quantity of output reach their peak early in life—in almost all fields before the age of 40. But he does note (p. 330) that " as a result of positive transfer the old usually possess greater wisdom and erudition. These are invaluable assets." Wisdom may well be the fruit of the contemplative years which lie beyond the point of " decreased sensitivity to environmental influences "—but by its very nature seems likely to prove resistant to objective measurement. This being the case, it is only natural that most of the quantitative findings so far presented in the literature, being of a kind dependent on continued interaction between the organism and the environment, should tend to show varying degrees of decline with age. On the available evidence from medical and physiological sources, Lehman (1953, p. 328) is probably justified in stating that " a decline occurs prior to 40 in physical vigour, energy and resistance to fatigue ". Adverse changes with age are also characteristic of the two principal distance receptors—the eye and the ear. For practical purposes it would appear desirable to face the fact, however unpalatable it may be emotionally, that in Lehman's words " possibly every human behaviour has its period of prime ". This done, it may be easier to see the objectives for psychological research on the process of ageing as being in the words of Bromley (1956b), " to point the way towards reduced stress and greater achievement during the mature and declining years ". These are " applied objectives "; the approach to them must be made both by the direct route and by the more circuitous path of relatively " pure " enquiry.

Perception

Reference has already been made to the eye and the ear: lest their pre-eminence be questioned, we may do well to note the judgment of Sherrington, as quoted by Adrian (1949): " The distance receptors are the great inaugurators of reaction . . . (they) integrate the individual not merely because of the wide ramification of their arcs to the effector organs through the lower motor centres; they integrate especially because of their great connections in the high cerebral centres."

Loss of elasticity in the lens starts in childhood, resulting in a gradual narrowing of the range of accommodation, while acuity for distant objects begins to show decline from about 45 years of age (Crouch, 19; 45 Dennis, 1953). Adequate knowledge of changes in the perception of depth and of colour is not yet available, but McFarland and Fisher (1955) have shown that complete dark adaptation, i.e., the final level as distinct from the rate, decreases very significantly with age. In their carefully controlled study the relationship was sufficiently close ($r+0{\cdot}9$) to justify the suggestion that the final threshold level might be useful as a functional measure of age. Kumnick (1956) has demonstrated substantial and linear relationships between age and pupillary responses to light and sound stimula " as reflected in decrease in mean maxima and minima pupil diameters, extent of constriction, and response velocity ". After drawing attention to the need for studies of such variables as states of stress, tension and fatigue, and the effects of various drugs, she adds the positive note that " the older eye apparently does not react as feebly as is generally assumed ". Kornzweig (1954), after listing the known and suspected effects of age on vision, also strikes a cheerful note by stating that " . . . only fifteen per cent have sufficient visual impairment to interfere with their activities . . . of this number many may be helped by surgery or by visual aids."

So far as hearing is concerned, it is well-established that there is a progressive hearing loss with age, principally however in the higher frequencies above 2,000 c.p.s. This results in problems of interpersonal communication, and although a good deal has been done on the remedial side through the development of small but efficient hearing aids, it does not seem as though any sound research has been done to evaluate the more general psychological consequences of this normal decrement. It is also necessary, as McFarland (1956) says, " to establish when, in the normal process of ageing, deterioration is no longer compensated by prosthetic aids, past experience, or a change in methods." To this last, one may well add the corollary that the possible value of " changes in method ", whether on the part of the subject or of those responsible for his immediate physical environment, should be a fruitful field of operations for the applied experimental psychologist. The effects of noise are not yet adequately understood: while it may be possible that the normal decremental process may be accelerated or its results made more severe by a sustained high-intensity noise in the working environment, it is known that a noisy background sometimes assists the older person to maintain an adequate communication level.

Finally, it is worth noting that no studies have yet been located in which *both* vision and hearing have been studied in the *same* individuals in relation to age, nor have the combined effects of both been taken into account when investigating, for example, the effects of increasing the sensory stimulus on motor performance—a subject to which we may now turn.

Motor Performance

The development of muscular strength, according to Fisher and Birren (1947) " follows a systematic trend with an increase up to the late twenties and a decline, usually at an increasing rate, from that time on ". Simple reaction time in response to an auditory stimulus has received recent attention in connection with ageing from Obrist (1953) and from Birren and Botwinick (1955), but nothing comparable for visual stimuli or contrasting or combining auditory stimulation has yet come to the notice of the writer. Apart from the apparent differences in experimental technique, and the fact that the proportionate increase for Birren and Botwinick's study is double that for Obrist's, the two investigations agree in providing evidence of a slowing down with age. Birren (1955) elsewhere quotes Rashevsky as his authority for stating that " in general, the stronger the peripheral stimulus, the shorter the reaction time ". The relation of this general observation to age might well repay close experimental examination.

Movement time was until recently also regarded as showing a marked increase with age, largely due to an early study by Miles (1931). However, from evidence presented by Leonard (1953), Szafran (1951) and Singleton (1954), consistent with the general findings of Rubin, von Trebra and Smith (1952), it seems clear that although there is some increase (mainly after 45–50 years of age) it is small in relation to that which has been demonstrated for decision making at the critical points in a psychomotor task. This leads us directly to the studies of skilled performance in relation to age carried out at Cambridge by Welford and his colleagues during the ten-year research period (1946–1956) supported by the Nuffield Foundation. Welford, following Bartlett, takes the view that " the peripheral receptor and effector organs are probably of only minor importance " so far as complex psychomotor skill is concerned. On the basis of the researches carried out by the Nuffield Unit, it is stated that the major causes of declining performance with age result from " failure to achieve some *fresh organization* of incoming data or outgoing action ", and of the two latter it is the failure in " perceptual organization " of the incoming data which is the more serious problem. Commenting on the Nuffield Unit's work in connection with this emphasis, Kleemeier (1954) draws special attention to the work of Szafran (1951) and makes the point that to explain his results (obtained in two very satisfactory experiments on aiming) one must conclude " either that older persons are reluctant to rely on reduced sensory cues in the performance of a task, or that they need to marshal all possible sensory data in task performance ". As the latter suggestion itself provides an explanation for the former, it is logically to be preferred, and it is consistent with Welford's theoretical position. Welford himself mentions (1953) the findings of Weston (1949*a* and *b*) that " poor lighting, small size of detail to be seen, and lack of contrast between objects viewed and their background can lower speed of work between the twenties and the fifties, even when subjects are matched for visual acuity by Snellen charts and Jaeger test types ". Improvement in one or other of these respects would of course increase the available sensory data relevant to the task, and this would consequently benefit the older people relatively more than the younger. Welford makes the point (1953) that analogous results might be obtained in an auditory inspection task where a particular sound might be critical.

Adult Learning

It is of course but a short step from considerations of sensory and psychomotor efficiency to an examination of our present knowledge about adult learning in relation to age. There are, broadly speaking, five main aspects of this fundamental psychological topic: intellectual ability, short-term retention of new material, long-term remembering, problem-solving, and creative thinking. This is not of course the only or even the best way in which to approach this general field of interest, and it may well prove desirable that others should advocate and illustrate an alternative ordering.

The most frequently quoted data come from three studies by Miles (1931); Jones and Conrad; and Wechsler (1955) (cf., e.g., Kaplan, 1945). These combine to show that performance on intelligence tests improves up to a point between 20–25 years of age, and then falls off steadily. Taking the Wechsler-Bellevue test as an example, it has long been known that certain sub-tests " hold-up " better with age than do others: Howell (1955) quotes five sources as agreed in demonstrating that these are consistently vocabulary, information and comprehension, while the most vulnerable are the picture arrangement, block design and digit symbol tests. Wechsler (1955) has recently provided fresh evidence to show that, contrary to widespread belief, " older subjects were negligibly, if at all, penalized by a speed factor, at least so far as the tests of the Wechsler Adult Scale are concerned ". This evidence took the form of comparing the performance of the same subjects under timed and untimed conditions of administration. Confirmation for this may be deduced from the demonstration by Vincent (n.d. in his *Age and Test Performance*) that a variety of timed and untimed tests, verbal and non-verbal, all showed a remarkably similar rate of decline with age, of the order of .027 standard deviation per year from 21–25 onwards.

Gilbert (1952) has shown that, with tested intelligence held constant, subjects in their sixties are considerably impaired in speed of learning and capacity for retention of new verbal material. McFarland (1956) however comments that " although it may be true that an average memory score is poorer at 65 than at 50 years, it is not known whether a loss in memory ability during this period consists of a small loss experienced by many, or of a great loss suffered by a few." His comment is no doubt prompted by Gilbert's own reference to the fact that among the brightest five per cent, the memory loss was only half as great.

The evidence concerning long-term remembering is conflicting, largely due to the almost complete absence of longitudinal studies of individuals and to reliance on material obtained from psychiatric subjects. Jones and Kaplan (1945) present a good deal of this conflicting material before coming to the conclusion that " since memory defects are frequently progressive, we are not justified in talking about senile memory as a single condition. The fact remains that memory defects are, perhaps, the most salient psychological symptoms of ageing, both normal and pathological." This seems rather strong for the available objective evidence.

So far as problem-solving is concerned, Clay (1954) has demonstrated that when time-stress is not present, the performance of older subjects (55+) compares well with that of younger subjects (<25) on a simple problem, but as the complexity of the problem is progressively increased, the older subjects become slower and less accurate, and finally in some cases fail to complete the task. Kay (1951) approached the study of problem-solving from an unusual angle by, in effect, giving his subjects the solution: their task was then to think how to make use of it. He says rightly that this technique " does give us some precision which is open to further analysis. By predicting certain types of mistake it is possible to confirm what it is that is holding up a subject's progress ". This approach enabled Kay to combine the results obtained in this task with those from a rote-learning study, leading him to the conclusion that it was " reluctance to discard their own ideas which characterized a poor performance both in the learning and in (the problem-solving) experiment ". The implications of this conclusion for the re-training of older workers must be obvious, and it also suggests the need for fresh examination of the oft-quoted " rigidity " of older people and the assumption that this is a " personality " feature of old age.

Further reference to creative thinking in the ordinary sense covered by the work of Lehman (already quoted) seems unnecessary, but attempts to tackle the problem directly by a planned study are so rare that a recent paper by Bromley (1956a) is most welcome. He used the Shaw Test, publicized by Howson (1948) and adapted by Bromley himself. It consists of four wooden blocks which can be arranged in different series, according to height, weight, position of a notch, and so on. Bromley concludes from the use of the test with 256 men and women, aged 17–82 years, that " although loss of intellectual efficiency with age does not entirely account for the decline in quantity and quality of creative output, it does account for the major part ". His measure of " intellectual efficiency " is the Wechsler-Bellevue Efficiency Quotient (WBEQ); this is defined (Wechsler, 1944, p. 220) as the mental ability score on the full-scale W-B " when compared with the score of the average individual 20–24 years of age ". This use of the term " creativity " is provisional, having in large part the connotation of " productivity " in circumstances where the subject's responses are for all practical purposes not limited by the nature of the task. In view of the work of Howell (1955) previously cited, it would be of great interest to see to what extent variations in " creativity " or " productivity ", as measured by the Shaw Test, can be accounted for in terms of the " hold " and " don't hold " groups of W-B sub-tests.

Emotional Stability

So far as it is possible to judge from the available literature, our firm knowledge on this topic is negligible. For some reason or other most of the studies reported have been based on the Rorschach test, and summary statements of this kind to be found in Gilbert (1952) arise from clinical impressions of highly self-selected subjects. There is clearly a great need here for sound research by relatively objective methods.

Work Performance

We come finally to a consideration of what is known about the effects of all the foregoing material on occupational performance as related to age. Welford (1955) points out that in this connection there is evidence for a series of changes in performance: " First, the original performance is maintained, but only by means of increased effort. Second, a change of method may enable achievement to be maintained or even improved. Finally, some fall of achievement occurs . . ." There is no doubt, both from laboratory studies and from industry itself, that " paced " (or " time-stressed ") tasks impose a burden great enough to handicap the middle-aged or elderly worker. The difficulties in learning to which we have referred do militate against the re-training of the older worker. Industrial reports, however, exemplified by that of Bowers (1952) stress the value of the older worker in terms of attendance, steadiness, conscientiousness, and accuracy. Some firms have set up special workshops for their oldest workers, and have shown that these need not be run without regard to economic considerations. (National Advisory Committee on the Employment of Older Men and Women, 1955). One of the big " unknowns " so far as the effects of age are concerned is surely the attitude of all concerned to the phenomenon. Tuckman and Lorge, quoted by Abrams (1955) provide some indication of the importance of this topic by reporting that " industrial managers and executives generally have a dim view of older workers, and are also filled with common stereotypes about them. Union leaders, they found, feel older workers are capable workers and are also victimized by prejudices of their own about older workers. Only social workers and government officials tended to view older workers in terms of individual differences ". Stereotypes about the rôle of the older worker are rife but rarely checked for their justification. Belbin (1955) however examined the commonly-held view that older people should be transferred to lighter work, and the implication that heavier work is not suited to older persons. He found that " those occupations which could be recognized as heavy and strenuous tended to have a higher proportion of persons engaged on them between the ages of 45 and 60 than did lighter skilled operations in the same industries ".

Conclusion

At various points in this survey the opportunity has been taken to note the desirability of further and better research. That there is much to be done needs no emphasis; what should be emphasized is that significant progress cannot be achieved without carefully-controlled longitudinal studies of individuals.

References

Abrams, A. J. 1955. " Discrimination against older workers in various countries." In *Old Age in the Modern World.* London: Livingstone, 291–295.

Adiran, E. D. 1949. *Sensory Integration.* Liverpool: Univ. of Liverpool Press, 20 pp.

Arnhoff, F. H. 1955. " Research problems in gerontology." *J. Geront.*, **10,** 452–456.

Beach, F. A. 1954. " The individual from conception to conceptualization." In Patton, R. A. (Ed.) *Current Trends in Psychology and the Behavioral Sciences.* Pittsburgh: Univ. of Pittsburgh Press, 82–114.

Belbin, R. M. 1955. " Older people and heavy work." *Brit. J. industr. Med.*, **12,** 309–319.

Birren, J. E. 1955. " Age changes in speed of simple responses and perception and their significance for complex behaviour." In *Old Age in the Modern World.* London: Livingstone, 235–247.

Birren, J. E. and Botwinick, J. 1955. " Age differences in finger, jaw, and foot reaction time to auditory stimuli." *J. Geront.*, **10,** 429–432.

Bowers, W. H. 1952. " An appraisal of worker characteristics as related to age." *J. appl. Psychol.*, **36,** 296–300.

Bromley, D. B. 1956a. " Some experimental tests of the effect of age on creative intellectual output." *J. Geront,* **11,** 74–82.

Bromley, D. B. 1956b. " Research prospects in the psychology of ageing." *J. ment. Sci.*, **102,** 272–279.

Clay, Hilary M. 1954. " Changes of performance with age on similar tasks of varying complexity." *Brit. J. Psychol.*, **45,** 7–13.

Crouch, E. L. 1945. " The relation between illumination and vision." *Illum. Engr. N.Y.*, **40,** 747–784.

Dennis, W. 1953. *Age and behaviour: a survey of the literature.* U.S.A.F. School of Aviation Medicine, Project No. 21–0202–0005, Report No. 1.

Fisher, M. B. and Birren, J. E. 1947. " Age and strength." *J. appl. Psychol.*, **31,** 490–497.

Gilbert, Jeanne G. 1952. *Understanding Old Age*. New York: Ronald Press.

Howell, R. J. 1955. "Sex differences and educational influences on a mental deterioration scale." *J. Geront.*, **10,** 190–193.

Howson, J. D. 1948. "Intellectual impairment associated with brain-injured patients as revealed by the Shaw test of abstract thought." *Canad. J. Psychol.*, **2,** 125–133.

Jones, H. E. and Kaplan, O. J. 1945. "Psychological aspects of mental disorders in later life." In O. J. Kaplan (Ed.) *Mental Disorders in Later Life*. Stanford, Calif.: Stanford Univ. Press, 69–115.

Kaplan, O. J. (Ed.). 1945. *Mental Disorders in Later Life*. Stanford, Calif.: Stanford Univ. Press.

Kay, H. 1951. "Learning of a serial task by different age groups." *Quart. J. exp. Psychol.*, **3,** 166–183.

Kleemeier, R. W. 1954. "Age changes in psychomotor capacity and productivity." *J. Bus. Univ. Chic.*, **27,** 146–155.

Kornzweig, A. L. 1954. "Physiological effects of age on the visual process." *Sight Sav. Rev.*, **24,** 130–138.

Kumnick, Lilian S. 1956. "Ageing and pupillary response to light and sound stimuli." *J. Geront.*, **11,** 38–45.

Lehman, H. C. 1953. *Age and Achievement*. New Jersey: Princeton Univ. Press.

Leonard, J. A. 1953. "Advance information in sensori-motor skills." *Quart J. exp. Psychol.*, **5,** 141–149.

McFarland, R. A. 1956. "The psychological aspects of ageing." *Bull. N.Y. Acad. Med.*, **32,** 14–32.

McFarland, R. A. and Fisher, M. B. 1955. "Alterations in dark adaptation as a function of age." *J. Geront.*, **10,** 424–428.

Miles, W. R. 1931. "Measures of certain abilities throughout the life span." *Proc. nat. Acad. Sci. Wash.*, **17,** 627–633.

National Advisory Committee on the Employment of Older Men and Women. Second Report. 1955. London: H.M. Stationery Office.

Obrist, W. D. 1953. "Simple auditory reaction time in aged adults." *J. Psychol.*, **35,** 259–266.

Rubin, R., von Trebra, P. and Smith, K. V. 1952. "Dimensional analysis of motion. III. Complexity of movement of pattern." *J. appl. Psychol.*, **36,** 272–276.

Singleton, W. T. 1954. "The change of movement timing with age." *Brit. J. Psychol.*, **45,** 166–172.

Szafran, J. 1951. "Changes with age and with exclusion of vision in performance at an aiming task." *Quart. J. exp. Psychol.*, **3,** 111–118.

Vincent, D. F. n.d. *Age and Test Performance*. London: National Institute of Industrial Psychology.

Wechsler, D. 1944. *The Measurement of Adult Intelligence*. Baltimore: Williams & Wilkins.

Wechsler, D. 1955. "The measurement and evaluation of intelligence of older persons." In *Old Age in the Modern World*. London: Livingstone, 275–279.

Welford, A. T 1953. "Extending the employment of older people." *Brit. med. J.*, **2,** 1193.

Welford, A. T. 1955. "Problems and methods of further research." In *Old Age in the Modern World*. London: Livingstone, 333–338.

Weston, H. C. 1949. "On age and illumination in relation to visual performance." *Trans. Illum. Engng. Soc.*, **14,** 281–297.

Weston, H. C. 1949*b*. *Sight, Light, and Efficiency*. London: Lewis.

DISCUSSION

F. Verzar. I admired the excellent lecture of Dr. Heron and should like to underline two points which he raised.

First, the importance of experimental psychological studies of the ageing process. He pointed out that there are not only deteriorations but also some well-proved improvements in certain qualities in the aged.

The irreversible decrease of elasticity of tendons and skin, the loss of physical strength, etc., are parts of the basic changes in the tissues during life. Our main practical aim in gerontology is, I think, not rejuvenation in the common sense, or prolongation of life beyond its natural limits, but rather to keep the aged in such physical health that they can use what we may summarize as their experience for the benefit of the community. Loss of physical health reduces life expectancy, but increased experience coupled with relative health makes the old person a leader of his group. From such a point of view, experimental psychological research is not only of theoretical but also of the highest practical value. It may be able to show us, not merely in general but also in certain definite cases, such as the work of factory workers, medical men, scientific workers, teachers, etc., the particular type of work in which the older individuals should replace the younger to the advantage both of themselves and of society.

I fully agree also with the second point which Dr. Heron mentioned. Research on ageing should not only be comparative, between groups of different ages, but should follow the ageing process in the same individuals. Two years ago a team was created in the University of Basle which has started work on such a research programme of long term investigations on the same individuals. Workers from a particular factory, in which they usually work from early youth to old age, are examined individually systematically every one or two years by an ophthalmologist for changes leading to presbyopia, by a medical officer for general health, by a physiologist for pulse wave velocity, etc. The first report was given in April 1956, by Dr. R. Bruckner at the Basle Symposium on Experimental Research on Ageing. We very much hope that similar research will also be started by other research groups.

NUTRITION AND AGEING

By

H. M. Sinclair

Laboratory of Human Nutrition, University of Oxford

Introduction

"Ageing" means becoming old or includes variants on this theme. For instance, Medawar (1955) defines it within a few lines as both "merely growing older" and "merely growing old". It is not clear that the concept of growth here adds anything other than confusion, and whereas becoming older implies mere passage of time, becoming old implies showing signs of old age. An infant as it loses its lanugo is becoming older and is showing signs of advancing age but not of old age; it would be unusual to say that the infant is showing signs of ageing. If we take ageing to mean becoming older, then an infant's expectation of life increases with ageing. If we take ageing to mean becoming old in the sense of showing signs of old age, then an infant cannot be said to be ageing except when it is in the condition of progeria. The process of acquiring the signs of old age is senescence, and the state reached when such signs are present is senility. I propose to discuss the effect of nutrition upon the process of senescence, and the effect of senility upon the process of nutrition.

Nutrition before Maturity

Maturation is the process of coming to full growth or development, which state is called maturity. There are various criteria of maturity: maximum length, ability to reproduce and the menarche are the three commonest. A fully-grown eunuch has reached maturity; a spawning fish has also reached maturity: therefore neither ability to reproduce nor maximum length can be taken as an absolute criterion.

A great many experiments have shown that undernourishment before maturity delays maturity and prolongs life. Either or both of these facts have been shown for silkworms (Kellogg and Bell, 1903), for butterflies (Pictet, 1905a, b), for *Tribolium confusum* (Chapman, 1920), for *Drosophila* (Northrup, 1917; Kopec, 1928), for *Daphnia* (Ingle, 1933; Ingle, Wood and Banta, 1937), for caterpillars of *Lymantria dispar* (Kopec, 1924), for tadpoles (Kopec, 1928), for the limpet *Patella vulgata* (Fischer-Piette, 1939), for

Protozoa (Rudzinska, 1952), for trout (McCay, Dilley and Crowell, 1929), for mice (Tannenbaum, 1947), for rats (Jackson, 1937; Saxton, 1945; Saxton and Silberberg, 1947; Riesen, Herbst, Walliker and Elvehjem, 1947; Templeton and Ershoff, 1949; Sherman, Campbell and Regan, 1949), and for cattle (Hansen and Steensberg, 1950). Northrup (1917) for example showed that undernourishment of *Drosophila* during the larval period prolonged this and increased duration of life from nineteen days to twenty-nine days. Kopec (1924) fed and starved caterpillars on alternate days, and thereby could prolong the duration of the larval stage from about 16 per cent to about 90 per cent of that of control animals, the prolongation being proportional to the degree of undernourishment; further, the pupal stage of the undernourished animals was shorter, and the duration of life of the imago was unaffected. Rudzinska (1952) showed that undernourishment of the sessile protozoan *Tokophrya infusionum* prolonged life whereas overnourishment shortened it. The most careful and extensive work has been done by McCay and his associates. For instance, in experiments started in 1930 (McCay, Sperling and Barnes, 1943; McCay, 1952) it was found that rats could be retarded as much as 1,150 days by undernourishment and still resume growth when adequately fed. They lived longer because of slower development of the chronic diseases that attacked the lungs, kidneys and middle ear, and tumours developed more slowly (as had been earlier shown in mice by Moreschi (1909)). McCay concluded that the type of aliments provided in an adequate basal diet mattered little in relation to life-span; the amount of aliments, or number of calories, was the important factor.

Is life prolonged through an effect that undernourishment before maturity has upon growth or upon development or upon metabolic rate? These three were dissociated by Gudernatsch's experiments (1912) with tadpoles. The simple explanation that rapid growth shortens life does not seem to be correct: Carlson and Hoelzel (1946) found that intermittent fasting prolonged the life-span of rats and retarded the development of mammary tumours, but without influencing the growth rate. McCay's undernourished rats were active, and although their basal metabolic rate was somewhat lowered their total metabolism was normal. The evidence is against the effects being brought about through altered rate of development. The effect of undernourishment is probably mediated through the pituitary. We could suppose that food stimulates the anterior pituitary to produce both gonadotrophic hormone, so that development takes place, and growth hormone which allows phospholipids to be synthesized in the body with the aid of essential fatty acids. If these phospholipids are formed, growth occurs; if they are not formed, through absence of essential fatty acids or absence of growth hormone, growth of the body does not occur.

Since it is very consistently found amongst lower animals that overnourishment hastens maturity and shortens the life-span whereas undernourishment delays maturity and prolongs life, the same might be true of man. In 1832 Edmonds published his "new theory of the cause producing health and longevity ". He believed that hardship in youth tended to decrease the rate of maturation and estimated that an increase of a year in the duration of infancy could increase the life-span by seven years. There is evidence that overnourishment of children hastens puberty (Bruch, 1941; Le Marquand, 1951). Conversely, undernourishment is known to delay puberty or the menarche (Wilson and Sutherland, 1953): we found this when we examined undernourished children in the Netherlands and Germany at the end of the war, and it would be interesting to follow these children, of whom we have records, through life. Brody (1945) showed that the proportion of life-span used in reaching maturity is constant for higher animals except man, who spends a relatively enormous proportion of his life in reaching puberty. For some years I have been almost alone in advising that by overfeeding children we may be harming them and shortening their life-span: " I think we should bear in mind that the optimum rate of growth of children is not necessarily the maximum, and that harm may be done by excessive feeding of children with milk or school meals and now by medication with vitamin B_{12} and aureomycin, even though these activities make them grow more quickly and mature earlier " (Sinclair, 1951). Even earlier (Sinclair, 1948) I had tried to point out that " insufficient thought has been given to the most desirable rate of growth, which is not necessarily the maximum rate. We can make a boy of 12 years taller and heavier than he would otherwise be by injecting anterior pituitary lobe extract; alternatively we can make him heavier and probably taller by superalimentation. There is indeed a tendency amongst nutritionists to regard the child of perfect nutriture as placid, rotund, red faced, and seated in contented contemplation of its folds of flesh. During the past several years there has been a marked increase in the rate of growth of children, although in England and the U.S. the adult male height has remained unchanged. It has not been shown that this increase in rate is necessarily advantageous; indeed it may be undesirable since the long time taken to reach maturity is characteristic of the human genus." This is a very important problem which was discussed at the Ciba Symposium last year (Sinclair, 1955a); my somewhat lonely crusade has been joined by McCance (1953; McCance and Widdowson, 1955). It may be that we should consider as the best diet in regard to aliments that which produces the longest life-span, and this is considerably less than many children are receiving. Maximum growth is not optimum growth. But it must be remembered that the object in man is to delay senility and increase the period of maturity before this develops; a long period of immaturity passing almost immediately into a long period of senility would be unfortunate. To some extent McCay's increased longevity was gained at the expense of the period of maturity before senility was apparent. The whole subject deserves more attention.

I have discussed aliments only because little is known about the effects that deficiency or excess of nutrients before maturity has upon senescence. For instance, Fritsch (1953) found that the amount of pantothenic acid in the medium was an important factor in determining the life-span of *Daphnia;* this and similar observations may complicate some of the results obtained with undernourishment. Lansing (1942) found that rotifers live longer in a medium low in calcium, and he suggests that senescence is caused by an increase in calcium at cell surfaces with lowered permeability and accumulation of toxic metabolites inside cells. Research in the field of nutrients is badly needed.

Nutrition after Maturity

Whereas most of our information about the effects that nutrition before maturity has upon senescence is derived from experiments upon lower animals, most of our information about the effects of nutrition after maturity is derived from observations upon man. There is good evidence that overnourishment after maturity increases the incidence of certain degenerative diseases.

In lower animals the evidence is conflicting. Though undernourishment of the larva of *Drosophila* increases life-span, undernourishment of the imago does not and may even decrease it (Kopek, 1928). However, adult ticks, *Dermacentor variabilis*, that have become attached to a host and are feeding freely, die within a few weeks, whereas unattached adults can live for more than two and a half years (Bishopp and Smith, 1938). McCay found that the effect of undernourishment upon life-span was much less in rats when it occurred after than when it occurred before maturity (McCay, Maynard, Sperling and Osgood, 1941).

It has been known from ancient times that fat men die earlier: *Plures crapula quam gladius* (Gluttony kills more than the sword). That careful observer Robert Burton may have come very near the truth when he wrote in the *Anatomy of Melancholy:* "As a lamp is choaked with a multitude of Oyl, or a little fire with overmuch wood quite extinguished; so is the natural heat with immoderate eating strangled in the body." Certain of those who had indulged immoderately in food and drink, such as Luigi Cornaro (1558) and the Miller of Essex, were driven to temperance by the dread of their former sufferings.

The early observations of the relation between overnourishment and early death have been strengthened by the statistics of the Life Insurance Companies, particularly those reported by Dublin (1930),

Dublin and Marks (1951) and Armstrong, Dublin, Wheatley and Marks (1951). Dublin analysed the mortality of persons who had been insured as sub-standard risks because of overweight between 1925 and 1934 and who were traced to 1950. It was high, and increased markedly with the degree of overweight. As I have pointed out to the Third Congress of the International Association of Gerontology (Sinclair, 1955b), " The chances of dying of cancer of the lung if one smokes over twenty-five cigarettes daily are smaller than the mortality risk from being a stone overweight." We may pass by the relatively minor disorders that are more frequent in the obese mainly for mechanical reasons; degenerative arthritis in knees, hips and lumbar spine; fractures; varicose veins; hernias; postural changes resulting in emphysema and diminished vital capacity; accidents. The fact that the obese are in general more clumsy than lean persons of the same age and are therefore less well equipped for avoiding motor-cars has nothing to do with senescence unless we assume that an increase in body-fat is part of the process of senescence. This increase appears to have a simple explanation. Muscular power and co-ordination are maximum at about the age of twenty-five years, and their decrease thereafter is certainly a criterion of senescence. It appears that appetite does not decrease as much as does muscular activity after this age, and of course there are usually inducements to eat more—hence the middle-aged spread. Unfortunately, increasing activity, as by a couple of rounds of golf at a week-end, does not necessarily check obesity and may actually increase it by stimulating appetite. The gestational complications of obesity are not entirely mechanical because, although obese mothers have large babies with great difficulty in obstetrical delivery, it appears that toxaemia is commoner in obese women and rarer in undernourished women (Chesley, Somers and Vann, 1948; Smith, 1947). It may be said that this is irrelevant to senescence; however, the older the woman at childbirth the greater are her chances of being obese.

The most interesting and important of the disorders that are commoner in the obese are the degenerative cardiovascular and renal diseases, diabetes mellitus, cirrhosis of the liver and cholelithiasis. We are only too familiar with the fact that these diseases also occur in persons who are not obese, and the incidence of all of them increases with age. They may be classed as nutritional disorders. In view of the rapidly increasing mortality in this country from cardiovascular disease and its relation to nutrition, some discussion is required.

I have put forward elsewhere reasons for believing that a relative deficiency of the essential fatty acids, linoleic and arachidonic, is becoming increasingly prevalent in this country as in most other highly civilized countries (Sinclair, 1956a, c, d). These acids are of course easily destroyed by oxidation, by hydrogenation and even by conjugation of the double bonds as by ultra-violet light. The "unnatural" (all-*trans* or *cis-trans*) isomers, and isomers with conjugated double bonds act in the body as antagonists of the ordinary required isomers which are all-*cis*. "Unnatural" is not an ideal term because cow's milk and butter contain almost none of the ordinary form of linoleic acid and a relatively large amount of *cis-trans* linoleic; cod-liver oil contains no ordinary linoleic acid. This may be yet another reason for thinking twice about pouring jugfuls of milk down the throats of children and spoonfuls of cod-liver oil down the throats of infants. Deficiency of essential fatty acids caused by consumption of diets low in them and containing the "unnatural" isomers produces disorders. High unsaturated fatty acids occur in the body in cholesteryl esters, in butyl alcohol, in glycerophosphatides (particularly ethanolamine phosphatide) and in cardiolipins (where linoleic acid represents five-sixths of the total fatty acids). These polyethenoid fatty acids include others besides linoleic: C_{20}— and C_{22}-polyethenoid acids occur abundantly in the glycerophosphatides of liver and brain (Klenk and Lindlar, 1955; Klenk and Dreike, 1955), and can be formed from linoleic in presence of vitamin B_6 (Witten and Holman, 1952) by addition of acetate followed by dehydrogenation (Steinberg, Slaton, Howton and Mead, 1956). This vitamin is not the only one that plays a part in their metabolism: vitamin E protects them from oxidation. But much of the vitamin E in our diets is destroyed by the so-called flour improvers; and the content of vitamin B_6, formerly not high, has decreased considerably since the bread subsidy ended on September 30, 1956.

When the essential fatty acids are absent they may be replaced by oleic acid or saturated fatty acids synthesized in the body; this may also occur if there is a relative excess of saturated fatty acids as perhaps on diets high in aliments, or in diabetes mellitus in which large amounts of fat are metabolized. The essential fatty acids can also be replaced by unusual highly unsaturated fatty acids if these are present in relative excess. The unusual compounds so formed do not function properly in the body and therefore can give rise to degenerative disease. For instance, Dr. Basnayake and I found two years ago that rats on a fat-free diet esterified cholesteryl with fatty acids synthesized in the body rather than with highly unsaturated fatty acids which the body cannot form, and the unusual ester was deposited in the skin; similarly, serum phosphatides were raised and had an unusually low iodine number. In coronary thrombosis the atheroma might be caused by deposition of an abnormal cholesteryl ester, and the increased coagulability of the blood might be caused by a raised concentration of ethanolamine phosphatide, which is known normally to contain an unusually unsaturated fatty acid and which Robinson and Poole (1956) have shown to increase the coagulability of blood.

There are two conditions in infants in which atheroma, commonly regarded as a sign of senescence, is said to occur: progeria and idiopathic hypercalcaemia. In the latter Bonham Carter has attributed the systolic murmur to hypertension and the hypertension to atheroma (Bonham Carter, Dent, Fowler and Harper, 1955), and I have given reasons for believing that the condition may be caused by deficiency of essential fatty acids and consequent increased sensitivity to vitamin D (Sinclair, 1956b); the adult form of this deficiency with adequate dietary calcium and vitamin D might be Mönckeberg's sclerosis. It is obscure why infantile progeria, which is premature senility of pituitary origin, should be accompanied by atheroma unless this is intimately associated with senescence.

I believe that essential fatty acids are in fact intimately associated with senescence. Possibly their most important function is to take part by polymerizing with protein in the formation of cell membranes and mitochondrial membranes. The greatly increased permeability of skin that occurs in deficiency of essential fatty acids (Sinclair, 1952; Basnayake and Sinclair, 1954, 1955) is probably caused by defective membranes of the cells of the stratum granulosum. It is possible that the alteration in permeability of old cells might be related to imperfection of the lipoprotein polymer. In deficiency of essential fatty acids oxidative phosphorylation is uncoupled (Klein and Johnson, 1954). This may well be because of imperfection of mitochondrial membranes which are of course essential for coupled oxidative phosphorylation. In old cells mitochondria become fragmented and finally disappear; a substance that is probably phospholipid accumulates in cells, and lipid accumulates in old nervous tissue. Pigment of the ceroid type which accumulates in old cells is related to peroxide formation from unsaturated fatty acids. It would be interesting to study these changes in old cells in relation to essential fatty acids.

Effects of Senescence on Nutrition

Digestion and absorption become impaired in old age. Loss of teeth, diminished saliva, and diminished gastric acid contribute to this. Half a century ago Metchnikoff thought that senescence was caused by poisons derived from intestinal putrefaction, and heroic measures were taken in the interests of intestinal hygiene; now we venerate our intestinal flora.

Transportation and utilization of nutrients and oxygen are less efficient in the elderly and therefore the nutrition of cells may be impaired.

Requirements may be altered. The increase in body-fat and decrease in muscle that occur with ageing raise two important questions: first, what is the " normal " amount of fat and what represents obesity in the elderly? secondly, should the calorie requirements be decreased? From the work of Shock it seems probable that the oxygen uptake of functioning cells in old individuals is no different from that in young subjects under resting conditions. An increase in body-fat probably indicates that old people are over-eating, and this increase is undesirable. There is a conflict of opinion about calcium requirements of the elderly: osteoporosis is probably more related to failure of protein anabolism than to dietary calcium deficiency.

The most important effects of senescence upon nutrition are those associated with conditions that are all too often found in old persons: poverty, ignorance and apathy.

References

Armstrong, D. B., Dublin, L. I., Wheatley, G. M., and Marks, H. H. 1951. *J. Amer. med. Ass.*, **147,** 1007.
Basnayake, V. and Sinclair, H. M. 1954. *J. Physiol.*, **126,** 55P.
Basnayake, V. and Sinclair, H. M. 1955. *Proc. 2nd Internat. Congr. Lipids, Ghent.*
Bishopp, F. C. and Smith, C. N. 1938. *U.S. Dept. Agric., Circ.*, 478.
Bonham Carter, R. E., Dent, C. E., Fowler, D. I. and Harper, C. M. 1955. *Arch. Dis. Childh.*, **153,** 399.
Brody, S. 1945. *Bioenergetics and Growth.* New York: Reinhold.
Bruch, H. 1941. *J. Pediat.*, **19,** 365.
Carlson, A. J. and Hoelzel, F. 1946. *J. Nutr.*, **31,** 363.
Chapman, R. N. 1920. *Ann. ent. Soc. Amer.*, **13,** 176.
Chesley, L. C., Somers, W. H. and Vann, F. H. 1948. *Amer. J. Obstet. Gynec.*, **56,** 409.
Cornaro, L. 1558. *Trattato della vita sobria.* Padova: Perchacino.
Dublin, L. I. 1930. *Hum. Biol.*, **2,** 159.
Dublin, L. I. and Marks, H. H. 1951. 60th An. Mtg. Assoc. Life Insurance Med. Dir. of America.
Fischer-Piette, E. 1939. *J. Conch.*, **83,** 303.
Fritsch, R. H. 1953. *Z. wiss. Zool.*, **157**, 35.
Gudernatsch, J. F. 1912. *Arch. Entwmech. Org.*, **35,** 57.
Hansen, K. and Steensberg, V. 1950. *Forskelligt opdraettede Koers holdbarked og ydelse.* No. 246, Udgivet af Statens Hysdyrbrugsudvalg. Kovenhavn.
Ingle, L. 1933. *Science*, **78,** 511.
Ingle, L., Wood, T. R. and Banta, A. M. 1937. *J. exp. Zool.*, **76,** 235.
Jackson, C. M. 1937. *Anat. Rec.*, **68,** 371.
Kellogg, V. L. and Bell, R. G. 1903. *Science*, **18,** 741.
Klein, P. D. and Johnson, R. M. 1954. *J. biol. Chem.*, **211,** 103.
Klenk, E. and Dreike, A. 1955. *Hoppe-Seyl. Z.*, **300,** 113.
Klenk, E. and Lindlar, F. 1955. *Hoppe-Seyl. Z.*, **299,** 74.
Kopec, S. 1924. *Biol. Bull.*, **46,** 1.
Kopec, S. 1928. *J. exp. Biol.*, **5,** 204.
Lansing, A. I. 1942. *J. exp. Zool.*, **91,** 195.
Le Marquand, H. S. 1951. *Proc. R. Soc. Med.*, **44,** 458.
McCance, R. A. 1953. *Lancet*, **ii,** 658 and 739.
McCance, R. A. and Widdowson, E. M. 1955. *Ciba Found. Colloq. Ageing*, **1,** 186.
McCay, C. M. 1952. Cowdry's *Problems of Ageing*, 3rd ed., Ch. vi, Baltimore: Williams and Wilkins.
McCay, C. M., Dilley, W. E. and Crowell, M. F. 1929. *J. Nutr.*, **1,** 233.
McCay, C. M., Maynard, L. A., Sperling, G. and Osgood, H. S. 1941. *J. Nutr.*, **21,** 45.

McCay, C. M., Sperling, G. and Barnes, L. L. 1943. *Arch. Biochem.*, **2**, 469.
Medawar, P. B. 1955. *Ciba Found. Colloq. Ageing*, **1**, 4.
Moreschi, C. 1909. *Z. Immun. Forsch.*, **2**, 651.
Northrup, J. H. 1917. *J. biol. Chem.*, **32**, 123.
Pictet, A. 1905a. *Mem. Soc. phys. d'hist. nat.*, Genève, **35**, 46.
Pictet, A. 1905b. *C. R.* 6me. *Congr. internat. Zool., Berne*, 498, 507.
Riesen, W. H., Herbst., E. J., Walliker, C. and Elvehjem, C. A. 1947. *Amer. J. Physiol.*, **148**, 614.
Robinson, D. S. and Poole, J. C. F. 1956. *Quart. J. exp. Physiol.*, **41**, 36.
Rudzinska, M. A. 1952. *J. Geront.*, **7**, 544.
Saxton, J. A. jun. 1945. *Biol. Symp.*, **11**, 177.
Saxton, J. A. jun. and Silberberg, M. 1947. *Amer. J. Anat.*, **81**, 445.
Sherman, H. C., Campbell, H. L. and Ragan, M. S. 1949. *J. Nutr.*, **37**, 317.
Sinclair, H. M. 1948. *Vitam. and Horm.*, **6**, 101.
Sinclair, H. M. 1951. *New Engl. J. Med.*, **245**, 39.
Sinclair, H. M. 1952. *Symp. biochem. Soc.*, **9**, 80.
Sinclair, H. M. 1955a. *Ciba Found. Colloq. Ageing*, **1**, 194.
Sinclair, H. M. 1955b. In *Old Age in the Modern World*, p. 160, Edinburgh: Livingstone.
Sinclair, H. M. 1956a. *Lancet*, **i**, 381.
Sinclair, H. M. 1956b. *Lancet*, **ii**, 101.
Sinclair, H. M. 1956c. *Proc. Nutr. Soc.*, **15**, xviii.
Sinclair, H. M. 1956d. *Proc. 3rd Internat. Congr. Lipids, Brussels.*
Smith, C. A. 1947. *Amer. J. Obstet. Gynec.*, **53**, 599.
Steinberg, G., Slaton, W. H. jun., Howton, D. R. and Mead, J. F. 1956. *J. biol. Chem.*, **220**, 257.
Tannenbaum, A. 1947. *Ann. N.Y Acad. Sci.*, **49**, 6.
Templeton, H. A. and Ershoff, B. A. 1949. *Amer. J. Physiol.*, **159**, 33.
Wilson, D. C. and Sutherland, I. 1953. *Brit. med. J.*, **ii**, 607.
Witten, P. W. and Holman, R. T. 1952. *Arch. Biochem.*, **41**, 266.

DISCUSSION

G. P. L. Miles. I do not presume to argue with Dr. Sinclair on the nutritive value of cow's milk but I must point out that the suggestion that the calves would not do very well on it but for the grass they consumed from the day of birth is a fallacy. Under ordinary farming conditions calves do not get any grass until three months old, but the best illustration is that of pedigree bulls sold for such large sums. These animals are fed almost exclusively on milk to an age when their mothers have to stand on a table to allow sucking. This is not a matter of fancy but because mother's milk produces the finest beasts.

A. Piney. Among the chemical accompaniments of the process of ageing, there are peculiar and little understood alterations in the metabolism of fatty materials. The intensity of these alterations varies greatly from person to person, but in everybody there is more or less widespread infiltration of lipid material into the intima and sub-intima of many arteries: the pathological condition known as atherosclerosis. Now, this progressive infiltration is accompanied by changes in the composition of the blood serum, changes which are only partly elucidated, although enough is already known to indicate that, even if they do not start the degenerative atherosclerotic process, they contribute to its progress.

Over the last forty years, many attempts have been made to incriminate one or more single fatty constituents of the serum. Thus, it has been supposed that a high level of blood cholesterol can lead to atheroma; certainly feeding experiments in rabbits suggest that this is the case. Others have incriminated the lipids in general, and so on.

My own observations strongly suggest that deposition of the fatty materials in the walls of arteries is facilitated, or perhaps even directly caused, by disproportion of the main fatty substances in the blood. Normally, the ratio between cholesterol and phospholipids is 1 : 1, but, in association with active atherosclerosis, this lipolytic quotient is usually raised but may be lowered. Some shift of the quotient seems to be a part of the ageing process, so that the idea that atheroma itself is an ineluctable consequence of innumerable microtraumata throughout life is not wholly acceptable. Furthermore, there is now a good deal of evidence that the rate of deposition in the arterial walls can be slowed by causing the lipolytic quotient to revert to normal, and by reducing the abnormally high beta fraction of the lipo-proteins. It seems to follow that research into the metabolism of fatty substances may be the most fruitful line of attack in the struggle against senility.

I hope later to publish details of the effects of administering a purified lecithin-fraction from soya beans, together with other surface-active agents, but already it is possible to assert that, in most cases, the pathologically altered lipid-spectrum reverts to normal or nearly normal.

Miss H. Saxl. Recent research carried out at the University of Leeds by the Nuffield Gerontological Research Unit indicates that an anti-lipaemic factor is formed in tissues which are rich in mucopolysaccharides such as the connective tissues of the vascular system, the liver and the blood, by the interaction of the pancreatic enzyme elastase with certain components of the serum. The following mechanism has been suggested for the reaction and is shown on the diagram.

A. In the presence of elastase and especially the fraction which has elastomucase activity, the mucopolysaccharides are made soluble and unstable.

B. The serum inhibitor for the enzyme elastase is related to the x-globulin fraction. Its action is to control the enzyme in the formation of degradation products. This was estimated as a proteolytic function with biochemical methods and it also inhibits the removal of the soluble polysaccharides and makes them stable. Histological studies indicate that metachromatic substances become orthochromatic and that an absorption complex of the orthochromatic and elastica staining substances occurs which has the appearance of an amorphous elastin under the electron microscope.

C. As a result of the above interaction, a lipase is activated. Biochemical investigations indicate that the quantity of inhibition is directly proportional to the amount of lipolysis. The lipase degrades neutral lipids and hydrolyses ester linkages.

The anti-lipaemic factor can act both on the lipids of vascular tissue and on those of the serum as a " clearing factor ".

DIAGRAM SHOWING A PROPOSED MECHANISM FOR THE FORMATION OF AN ANTI-LIPAEMIC FACTOR

E, Elastin; C. Collagen; G.S., Ground Substance; S.P., Sulphated Polysaccharides; C.F., Clearing Factor; L.P., Lipoprotein; Sol., Soluble; —ve, Negative.

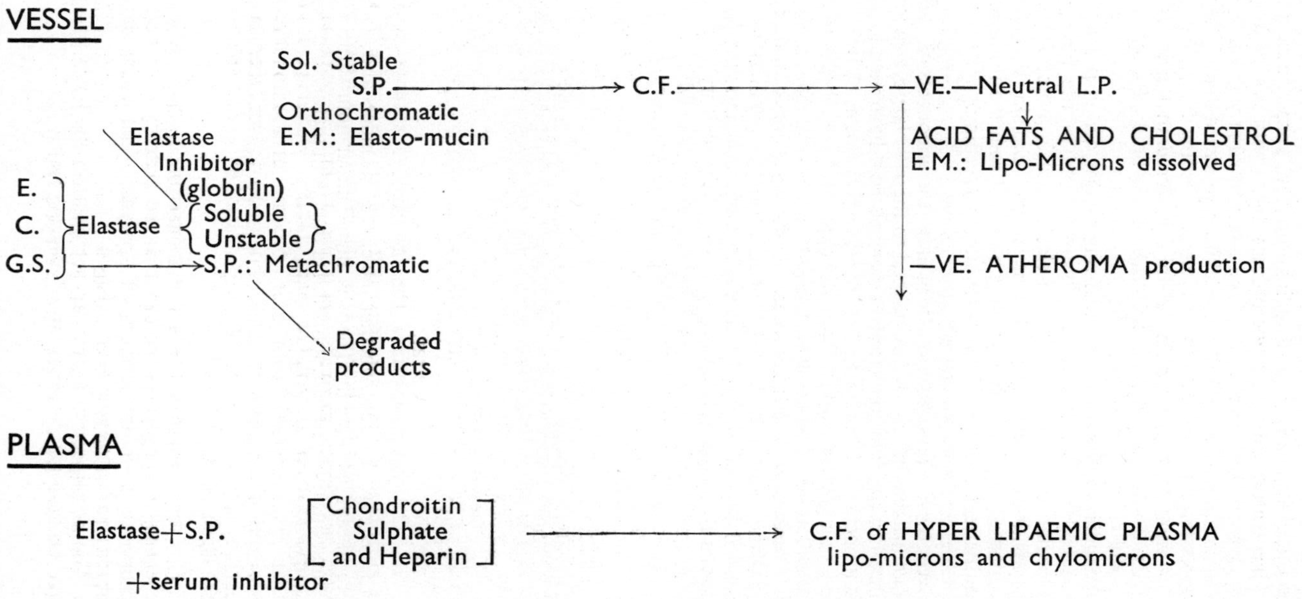

The classical discovery of Anitschkow, who produced a lesion simulating human atherosclerosis by feeding rabbits with cholesterol, has resulted in an intensive effort over the last twenty years to link atherosclerosis to a faulty lipid metabolism. Anitschkow observed that if the ingestion of cholesterol was discontinued after the feeding had been carried out for 3–4 months, a resorption of lipid occurred in the inner layers of the arterial wall, although the characteristic lipid plaque remained in the media for a much longer time.

Studies of the localization of the enzyme action point to the fact that the high concentration of polysaccharides in the inner layers of the vascular wall, together with an elastomucase, protects these against the deposition of the lipid. This would explain the observations of Anitschkow that the lipid was absorbed from the intima of the arterial wall after the cessation of the cholesterol feeding.

G. C. Kennedy. I wonder if it is too inconoclastic to question the significance that has been attached to these " classic " papers of McCay's that Dr. Sinclair and several other speakers have mentioned. It is true, is it not, that McCay had to make several attempts before he could get the housing conditions just right for the undernourished animals to survive at all? One wonders what the final life tables would have looked like if all these early deaths had been included. There is an independent study of some of McCay's work by Saxton and Kimball (1941. *Arch. Path.*, **32,** 951), which suggests that the effect of nutrition was different on each of the main diseases observed in rats. Respiratory infections were just as common in rats which were killed as in those which died, that is they did not seem to be a cause of death, and the incidence was unaffected by diet. Tumours were more common in the undernourished rats, which lived to the " tumour age ". Only the renal diseases, which are probably an important cause of death in rats, were ameliorated by undernutrition. These renal diseases have been studied by many different people, and their onset can be accelerated by a variety of methods. The simplest is merely to reduce the amount of renal tissue surgically. If this is done, renal breakdown occurs quite rapidly even in young rats, and is accompanied by loss of weight, staring fur and a general appearance such as is shown in McCay's pictures of " old " rats. I think it is simply the appearance of the chronic uraemic rat.

E. Digby. I doubt whether the effects of under- and over-nourishment are always as stated. There is much under-nourishment in India and China but the general impression is that both maturation and puberty are earlier there than in Western Europe.

G. H. Bourne. The answer to Mr. Digby is that maturity does not occur earlier in Eastern countries.

G. I. M. Swyer. There is no evidence that Indian women are nubile at an earlier age than Europeans.

F. Le Gros Clark. I know of no real racial differences in the maturation of human beings. There is some evidence that in Belgium, wartime under-nourishment led to a delay in the onset of puberty.

T. Levett. In his very helpful contribution Dr. Sinclair observed that tumours grew more slowly in instances of undernourishment. Does this apply to man or to the rat? If to the latter, how does this correlate with the observations that deprivation of iodine in the food of rats is conducive to the production of tumours in the thyroid gland?

Second, Dr. Sinclair showed a sudden rise of coronary thrombosis starting in 1926. Could this be a relative increase due to the use of the electrocardiogram in diagnosis which became popular at about this period?

H. M. Sinclair. The effect of the introduction of electrocardiograms was allowed for in the study on coronary thrombosis.

GENETIC VARIATIONS IN AGEING

By

J. MAYNARD SMITH

Department of Zoology, University College, London, W.C. 1

THERE are two kinds of differences between individuals which can form the starting point of a genetical investigation of ageing, namely differences in the length of life and in the cause of death. Of course these two things are related. A man carrying the sex-linked gene responsible for haemophilia has a lower expectation of life than normal, and is likely to die from excessive bleeding. After the entry of Europeans into Tierra del Fuego, the natives had a lower expectation of life than the immigrants, mainly because they could not survive an attack of measles, from which many of them died, although the children of mixed parentage survived (Bridges, 1948). However, although these two examples show that people of certain genotypes, in certain circumstances, have a lower expectation of life because they are susceptible to a particular cause of death, neither is strictly relevant to the problem of ageing, because in neither case does the susceptibility of an individual to the factor causing death increase with age.

The study of ageing is concerned with those changes in individuals with age which render them more liable to succumb to one or more possible causes of death. Therefore a genetical study of the differences in life span in any animal is possible only if the animals in question can be kept in conditions such that at least the majority of them die from causes which they would have survived had they been younger. To meet this condition, it is not necessary to be able to determine the cause of death in any particular case. It is sufficient to show that, in the given culture conditions, the force of mortality increases with age. Figure 1 shows adult survival curves for a genetically homogeneous population of the fruitfly, *Drosophila subobscura*, kept in mated pairs at 25°C., and transferred to fresh food every second day. In no case is it possible to state the cause of death of an individual. It is however quite clear that individuals become more likely to die, from whatever cause, as they grow older. Such a survival curve can be obtained only if conditions are uniform, and such that " accidental " deaths are rare; by an accidental death I mean a death from causes which would kill an individual at whatever age it was exposed to them.

Figure 1 also shows that life span is, in this and most other populations, a continuously varying character. Individuals do not fall into two or a few sharply defined classes. Such characters present special difficulties for genetical analysis for two reasons. First, the continuous distribution suggests that they are influenced by genes at many different loci, and second, they may be influenced by the environment as well as by genetic make-up. For example, the flies whose survival curve is shown in Figure 1 were kept as larvae and pupae at 25°C. Flies which were kept during pre-adult life at 15°C. survived for about 20 per cent longer as adults at 25°C.

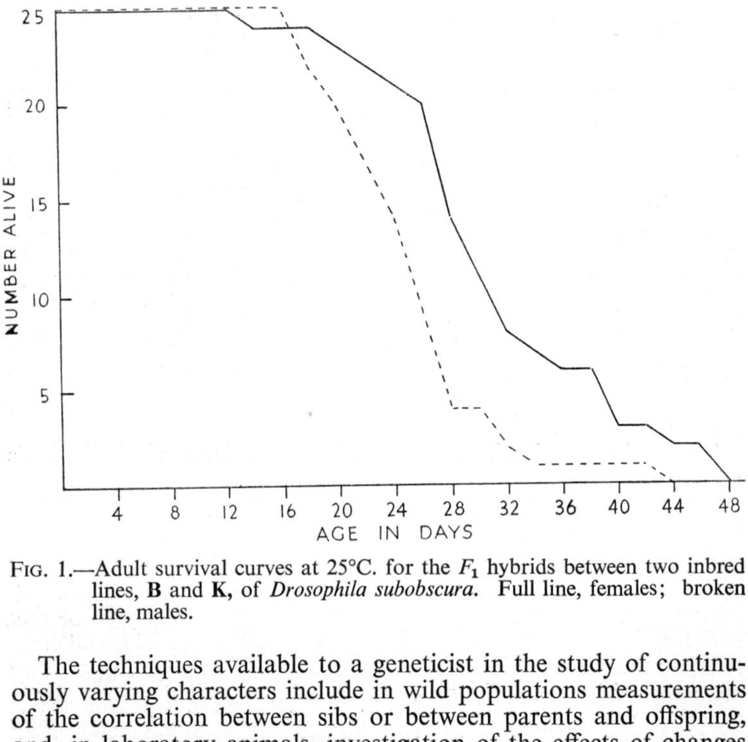

FIG. 1.—Adult survival curves at 25°C. for the F_1 hybrids between two inbred lines, **B** and **K,** of *Drosophila subobscura.* Full line, females; broken line, males.

The techniques available to a geneticist in the study of continuously varying characters include in wild populations measurements of the correlation between sibs or between parents and offspring, and, in laboratory animals, investigation of the effects of changes in the breeding system, for example by inbreeding or distant hybridization, and of selection. Adult life span is in many ways an inconvenient character for such studies, since it is not known until the individual is dead. Thus selection for a shorter adult life span could be carried out only by breeding from a large number of pairs set up at random, keeping the parents until they died of old age, and continuing breeding from the offspring of those parents which died young. Selection for increased life span is easier, in that individuals can be kept unmated until a certain proportion have died, and then the survivors used as parents of the next generation. There is however the danger in this procedure that maternal age itself may influence the longevity of the offspring. In parthenogenetically reproducing rotifers, Lansing (1948) showed that lines maintained from the eggs laid by young mothers could be kept indefinitely, whereas if the later eggs laid by a female were used, there was a progressive decline in mean life span in successive generations, and eventual extinction. It is not known how general such an effect will prove to be in sexually reproducing organisms, but some special cases are known. For example, mongoloid idiots have a low expectation of life, and are commoner among the children of old mothers (Penrose, 1954).

For these reasons, there is less information on the genetics of ageing than of many other quantitative characters. However, what information there is is of some interest. I shall describe first some experiments on laboratory populations of *Drosophila,* and then consider how far these agree with observations on human populations. The first genetical studies on ageing in *Drosophila melanogaster* were due to Pearl (1928). He found that a stock homozygous for the recessive mutant *vestigial* had an expectation of life about half that of a relatively outbred wild-type stock. The first generation hybrids, wild-type in appearance, lived for slightly longer than their wild-type parents, which suggests that crossing unrelated animals may give offspring with an increased life span. In the second generation, the wild-type flies resembled in life span their wild-type grandparents, showing that the advantages of hybridization were lost in the F_2; the vestigial flies resembled their vestigial grandparents, showing that a short life span is a feature of flies homozygous for this mutant. Other examples could be given in which individuals homozygous for a particular recessive gene with striking phenotypic effects live for a much shorter time than do normal individuals. It does not follow that there also exist recessive mutants for which homozygotes live appreciably longer than the rest of the population. Genetic mutations originate as more or less random changes in the biochemical properties of chromosomes; changes with large effects are likely to be deleterious. By analogy, large random changes in a motor car engine would be more likely to reduce than to increase its efficiency. It is possible that a single mutation which lowered the metabolic rate, or which delayed metamorphosis, might appreciably increase the life span. Normally, however, long life, in so far as it is genetically determined, is likely to be due to a favourable constellation of genes at a large number of loci.

An investigation of the effects of selection on adult life span was carried out by Comfort (1953) in *Drosophila subobscura.* In eight successive generations in a wild-type stock he bred only from those individuals which survived for more than 30 days, which was greater than the population mean. There was no increase in mean life span neither was there any decrease, as might have been expected if the effects found by Lansing in rotifers operate also in *Drosophila*. The absence of any response to selection might suggest that the differences in longevity were environmental and not genetic in origin. This however is not the case. Figure 2 shows survival curves for

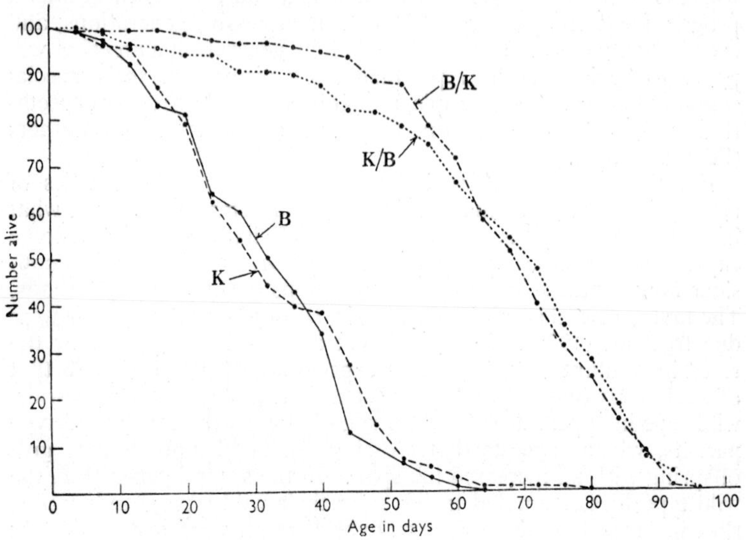

FIG. 2.—Adult survival curves (sexes combined) at 19° C. for two inbred lines, **B** and **K**, of *Drosophila subobscura*, and of the F_1 hybrids between them (Clarke & Maynard Smith, 1955).

two inbred lines of *D. subobscura*, and for the reciprocal hybrids between them, kept as far as possible in identical conditions. The mean expectation of life of the hybrids is approximately twice that of the inbred flies, and the difference is largely or wholly genetic. It will be recalled that Pearl also found that F_1 hybrids lived longer than their parents, but the difference in the present case was far greater, probably because the parental flies were highly inbred. It should not be thought that these results demonstrate that hybridization increases the life span above that normal for the species. The offspring of fertilized females caught in the wild lived in laboratory conditions for as long as did these hybrids, although there was a small proportion of very short-lived flies among the offspring of wild females. Thus the explanation of these results is that inbreeding greatly reduced the normal life span, which was restored by crossing the inbred lines.

The simplest explanation of these facts is that genetic heterozygotes live for longer than do homozygotes. Consider what would happen if differences in longevity were due to a pair of alleles, *A* and *a*, at a single locus, such that individuals with the genotype *Aa* lived for a long time, and *AA*, *aa* for a short time. First, selection would be ineffective; mating together two long-lived heterozygotes, *Aa* × *Aa*, would give offspring half of which were short-lived homozygotes, *AA* or *aa*, whereas mating together two short-lived flies might give nothing but short-lived offspring (if the mating were *AA* × *AA* or *aa* × *aa*), or nothing but long-lived offspring (if the mating were *AA* × *aa*). There would in fact be no correlation between parents and offspring, although there would be a correlation between sibs. In contrast, inbreeding would produce homozygous and uniformly short-lived flies. In practice, not one but many loci are involved. Different inbred lines tend to become homozygous for different alleles at many of these loci, so that hybridization restores both genetic heterozygosity and long life. The conclusions that selection would be ineffective and that inbreeding would reduce the life span continue to hold if there are many pairs of alleles for which the heterozygotes have a greater expectation of life than either homozygote. Such genes are said to show " heterosis ", in contrast to genes with " additive " effects, in which the heterozygote is intermediate between the homozygotes.

The pattern of inheritance typical for characters influenced by genes with heterotic effects has been found for rate of development as well as for longevity in *D. subobscura* (Hollingsworth and Maynard Smith, 1955). In contrast, the genetic variance of the number of bristles on the abdomen of *D. melanogaster* has been shown by Mather and Harrison (1949) and later workers to be due in the main to genes with additive effects, or showing simple dominance relationships; there is a response to selection either for increased or decreased bristle number, and the offspring of crosses between lines differing in bristle number tend to be intermediate between their parents. It is not an accident that the former characters, which are closely associated with fitness, should show heterosis, whereas bristle number should show in the main additive inheritance. Haldane (1949) showed that in a random mating population in equilibrium under the influence of natural selection, no correlation would be expected between the fitness of parents and offspring, although there will be a substantial correlation between sibs. The reason is as follows: two alleles at a given locus, *A* and *a*, will both be present in a population in high frequency only if the heterozygote *Aa* is fitter than either homozygote, since if one homozygote, say *AA*, is fitter than either *Aa* or *aa*, the allele *a* will be eliminated by selection, being maintained at a low frequency only by mutation. Therefore most of the genetic variance of fitness will be heterotic. It has already been pointed out that in this case there will be no parent-offspring correlation, and no response to selection. To put the matter in another way, if natural selection has already maximized fitness, it is impossible further to increase fitness by artificial selection. The same will be true of characters such as longevity which are closely associated with fitness, although it might be possible to increase the mean life span by selection at the expense of reduced fitness in other ways, for example reduced fertility.

There is some evidence that the genetic variance of longevity in man is in the main due to genes with heterotic effects. Beeton and Pearson (1901) measured parent-offspring and sib-sib correlations for human longevity. They studied Quaker families, to ensure that the measured individuals were as far as possible living in the same economic conditions, and therefore that any correlations obtained were caused by genetic rather than environmental factors. They considered only those children who died at 21 or later, on the grounds that the causes of death in infancy and in later life are different. The mean value of the four parent-offspring correlations was 0.1365, or only about one-quarter of that for human stature, and of the three sib-sib correlations 0.2611. The higher value of the latter correlation is what would be expected from Haldane's argument. Apparently it is not true that to live a long time you must choose long-lived parents.

The genotype of an individual may influence not only its life span but also the particular cause of death. This aspect of the genetics of ageing is most easily studied in mammals, since too little is known of the pathology of most other groups to determine the cause of death. There is little doubt that animals of similar genotypes, kept in conditions in which there are few accidental deaths, tend to die of similar causes. For example, inbred lines of mice differ not only in the age of onset of cancers, but also in the particular kinds of cancer which develop. Similarly, different breeds of domestic dogs tend to die from characteristic causes (e.g. Cotchin, 1954).

In man, there is a progressive rise of blood pressure with age, which may contribute to death from cardiac disease or stroke. Hamilton *et al* (1954) measured the arterial pressure in the relatives of patients with " normal " blood pressures, and of patients with " essential hypertension ", i.e. in patients with blood pressures higher than an arbitrarily selected value, and having no disease to which these pressures could be attributed. When allowance was made for differences due to age and sex, the blood pressures of the former group of relatives did not differ from the general population, whereas the pressures of relatives of patients with essential hypertension were higher, the excess for the relatives over the normal for their age and sex being about 20 per cent of that for the affected patients. This suggests that there is a genetic element in the determination of high blood pressure, although, as the authors emphasize, the blood pressure of individuals increases with age, and is probably influenced also by environmental factors.

These results have been quoted for two reasons. First, they confirm that genetic factors influence a particular process which may contribute to death in old age. Second, it was found that the resemblance between parents and their children was as great as that between sibs. This is in contrast to the results quoted earlier for total life span. The contrast is not really surprising. Haldane's conclusion that there will be no parent-offspring correlation holds only for overall fitness, although the correlation is likely to be small for characters closely associated with fitness. Such an association may exist between expectation of life and total fitness, but not between fitness and the ability to withstand one particular group of mortality factors. A high blood pressure may contribute to death from cardiac disease in old age, but cannot have consistently adverse effects on fitness, since if a high blood pressure were uniformly disadvantageous natural selection would reduce the mean level in the population. It is therefore probable that the deleterious effects of high blood pressure in old age are counterbalanced by advantages, perhaps earlier in life, natural selection maintaining the mean arterial pressure in the population at an optimal value. If so, we should expect to find a parent-offspring correlation, just as we do in the case of human stature.

To sum up, differences in longevity present special difficulties for genetic analysis, in particular because an individual cannot be classified until it is dead, and even then only if it has been kept in uniform and favourable conditions. There is no doubt that some differences in longevity are genetically determined, and that animals of similar genotypes tend to die of similar causes. Longevity, like other characters closely associated with fitness, tends to be a property of genetically heterozygous organisms. Consequently there is likely to be little correlation between the life spans of parents and offspring in random mating populations, though there will be a correlation between sibs; this conclusion has been confirmed for human populations. Other consequences of this kind of genetic determination, which have been confirmed in *Drosophila*, are that artificial selection will be relatively ineffective in increasing longevity, and that inbreeding, at least in naturally outbreeding species, drastically reduces the expectation of life, which can be restored by subsequent hybridization between inbred lines.

References

Beeton, M. and Pearson, K. 1901. " On the inheritance of the duration of life, and on the intensity of natural selection in man." *Biometrika*, **1**, 50–89.

Bridges, E. L. 1948. *Uttermost Part of the Earth.* London: Hodder & Stoughton.

Clarke, J. M. and Maynard Smith, J. 1955. " The genetics and cytology of *Drosophila subobscura.* XI. Hybrid vigour and longevity." *J. Genet.*, **53**, 172–180.

Comfort, A. 1953. " Absence of a Lansing effect in *Drosophila subobscura.*" *Nature, Lond.*, **172**, 83.

Cotchin, E. 1954. " Neoplasia in the dog." *Vet. Rec.*, **66**, 879–885.

Haldane, J. B. S. 1949. " Parental and fraternal correlations for fitness." *Ann. Eugen., Lond.*, **14**, 288–292.

Hamilton, M., Pickering, G. W., Fraser Roberts, J. A. and Sowry, G. S. C. 1954. " The aetiology of essential hypertension. IV. The role of inheritance." *Clin. Sci.*, **13**, 273–304.

Hollingsworth, M. J. and Maynard Smith, J. 1955. " The effects of inbreeding on rate of development and on fertility in *Drosophila subobscura*." *J. Genet.* **53**, 295–314.

Lansing, A. I. 1948. " Evidence for ageing as a consequence of growth cessation." *Proc. nat. Acad. Sci. Wash*, **34**, 304–310.

Mather, K. and Harrison, B. J. 1949. " The manifold effect of selection." *Heredity*, **3**, 1–52 and 131–162.

Pearl, R. 1928. *The Rate of Living*. University of London Press.

Penrose, L. S. 1954. " Mongolian idiocy (mongolism) and maternal age." *Ann. N.Y. Acad. Sci.*, **57**, 494.

DISCUSSION

N. W. Pirie. Was the group of people for whom was found a correlation of only 0.13 between longevity of the parents and longevity of the children, a random sample of the population? If not, may the low correlation, which runs counter to so much general opinion, be due to the community being already unusually long lived or in some other way abnormal?

J. Maynard Smith. The group chosen was not a random sample: it was drawn from Quaker families. The community was unlikely to have been abnormal since in the same population quite high correlation was found for other characters such as statute.

AUTHOR INDEX

AUTHOR INDEX

The following index refers to the pages on which contributions by persons named are to be found. It does not include the names of persons whose papers have been cited as references.

SUBJECT INDEX